SUSTAINABLE ENGINEERING PRACTICE

An Introduction

SPONSORED BY
The Committee on Sustainability of Technical Activities Committee

Published by the American Society of Civil Engineers

Cataloging-in-Publication Data

Sustainable engineering practice : an introduction / sponsored by the Committee on Sustainability of Technical Activities Committee.
    p. cm.
Includes bibliographical references.
ISBN 0-7844-0750-9
1. Environmental engineering. I. American Society of Civil Engineers. Committee on Sustainability.

  TA170.S87 2004
  628--dc22

                2004018323

Published by American Society of Civil Engineers
1801 Alexander Bell Drive
Reston, Virginia 20191
www.pubs.asce.org

Table of Contents

Preface

"Creating a sustainable world that provides a safe, secure, healthy life for all peoples is a priority for the US engineering community. It is evident that the US engineering must increase its focus on sharing and disseminating information, knowledge and technology that provides access to minerals, materials, energy, water, food and public health while addressing basic human needs, Engineers must deliver solutions that are technically viable, commercially feasible, and environmentally and socially sustainable."

This declaration comes from a "Dialogue on the Engineers Role in Sustainable Development – Johannesburg and Beyond" that was held at the National Academy of Engineering in Washington, D.C. on June 24, 2002 (NAE, 2002). The participants represented a "who's who" of professional engineering organizations, federal departments and agencies, and others with interests and responsibilities in sustainable development. The Declaration also committed its signatories "…to moving forward in support of the US engineering community to meet societal needs through capacity building, improved education, training, information development and dissemination, and engaging the engineering profession in all stages of the decision process."

The Committee on Sustainability of the Technical Activities Committee of ASCE prepared this Committee Report in part, as a response to the Declaration. The Report provides a starting point for engineering students and young engineering professionals in practice to gain an understanding of the basic principles of sustainability and their application to engineering work. It is intended to fill a need for a "primer" on sustainability that can be introduced early in an engineer's career: it brings together all the basic dimensions of its history, concepts, and applications; and through a variety of examples and references, can inspire and encourage engineers to pursue and integrate sustainable engineering into their work on life-long basis.

The Report also responds to a common issue, the achievement of development that is sustainable, which poses an opportunity and a challenge to the engineering profession, which according to Hatch (2002), are: "…opportunity for greater public service (and economic gain) and challenge to our traditional education, our methods, our technologies and even our ethics." Hatch further states that achievement of sustainable development will change how the public, clients, students, employees, and the youth that professions and industries seek to attract, perceive engineers. The relevancy and reputation of engineering will depend largely on the willingness and demonstrated contribution of the profession to achieving sustainability. This is a vision, an ethic, not a strategy and supporting tactics, not a set of specific technologies, processes, laws, regulations or standards.

The World Commission on Environment and Development took the lead in outlining a sustainable future, culminating in the 1987 publication of <u>Our Common Future</u> (WCED, 1987). This report (often referred to as the Brundtland Report) provided a focused definition for the concept that has endured: "[Sustainable Development]…is a process of change in which the exploitation of resources, the direction of investments, the orientation of technological development, and institutional change are all in harmony and enhance both current and future potential to meet human needs and aspirations." In other words, it is "Meeting the needs of the present without compromising the ability of future generations to meet their own needs."

According to Hatch (2002), this definition focuses on <u>environmentally</u> sustainable development. Over the years he has been adding other adverbial modifiers to the phrase "sustainable development" to better express the context of sustainability, beginning with the obvious one in a market economy – <u>economically</u> sustainable. In a free democratic society (or even in one that is not), he adds <u>politically</u> sustainable. Furthermore, in a socially conscious society, he adds <u>socially</u> <u>sustainable</u>. More recently, he also has added <u>ethically</u> sustainable as people have become more aware of the religious and ethical underpinnings of so much of human behavior – particularly the extreme. These adverbs are not independent but are clearly interdependent and any successful endeavor must address and satisfy them all.

The Report is structured as follows:

> *Chapter I - Background* begins with a challenging statement of a Pulitzer Prize-winning environmental scientist, as we enter what he calls "The Century of the Environment." It is followed by a summary of the role and accomplishments of engineers in sustainable development. (The complete report, "Engineers and Sustainable Development," is contained on a CD appendix to this report.) Chapter I concludes with a summary of the major commitments made and implementation activities agreed upon at the World Summit on Sustainable Development, held in Johannesburg, South Africa, in September 2002, and the initial steps being taken by the US engineering community and its global partners in moving "beyond Johannesburg." The definition of, and policy for, sustainability by the American Society of Civil Engineers is shown in Sidebar 1, at the end of Chapter I.

> *Chapter II - Selected Readings* offers a broad spectrum of examples, which describe how sustainability principles can and are being integrated and applied in engineering education, research, and practice.

Chapter III - Sustainability Definitions, Policy Statements, and Principles will give the reader a broad sense of the visionary and ethical goals and strategies of sustainable engineering practice.

Finally, *Appendix A - References, Resources, and Tools* contains a wealth of additional information that can be used to pursue the basic material in the Report in greater depth and detail.

Chapter I
Background

The Century of the Environment – A Challenge[1]

"The twentieth century was a time of exponential scientific and technical advance, the freeing of the arts by an exuberant modernism, and the spread of democracy and human rights throughout the world. It was also a dark and savage age of world wars, genocide and totalitarian ideologies that came dangerously close to global domination. While preoccupied with all this tumult, humanity managed collaterally to decimate the natural environment and draw down the nonrenewable resources of the planet with cheerful abandon. We thereby accelerated the erasure of entire ecosystems and the extinction of thousands of million-year-old species. If Earth's ability to support our growth is finite–and it is–we were mostly too busy to notice.

As a new century begins, we have begun to awaken from this delirium. Now, increasingly post-ideological in temper, we may be ready to settle down before we wreck the planet. It is time to sort out Earth and calculate what it will take to provide a satisfying and sustainable life for everyone into the indefinite future. The question of the century is: How best can we shift to a culture of permanence, both for ourselves and for the biosphere that sustains us?

The bottom line is different from that generally assumed by our leading economists and public philosophers. They have mostly ignored the numbers that count. Consider that with the global population past six billion and on its way to eight billion or more by mid-century, per-capita fresh water and arable land are descending to levels resource experts agree are risky. The ecological footprint—the average amount of productive land and shallow sea appropriated by each person in bits and pieces form around the world for food, water, housing, energy, transportation, commerce, and waste absorption—is about one hectare (2.5 acres) in developing nations but about 9.6 hectares (24 acres) in the United States. The footprint for the total human population is 2.1 hectares (5.2 acres). For every person in the world to reach present U.S. levels of consumption with existing technology would require four more planet Earths. The five billion people of the developing countries may never wish to attain this level of profligacy. But in trying to achieve at

---

[1] This section is an abstract from "The Future of Life" by E. O. Wilson (2002), Pulitzer Prize-winning author and environmental scientist, who sets out a challenge to engineers and scientists in the "Century of the Environment." This material is included in this report with permission.

least a decent standard of living, they have joined the industrial world in erasing the last of the natural environments. At the same time *Homo sapiens* has become a geophysical force, the first species in the history of the planet to attain that dubious distinction. We have driven atmospheric carbon dioxide to the highest levels in at least two hundred years, unbalanced the nitrogen cycle, and contributed to a global warming that will ultimately be bad news everywhere.

In short, we have entered the Century of the Environment, in which the immediate future is usefully conceived as a bottleneck. Science and technology, combined with a lack of self-understanding and a Paleolithic obstinacy, brought us to where we are today. Now science and technology, combined with foresight and moral courage, must see us through the bottleneck and out." (Wilson 2002)

Engineers and Sustainable Development[2]

Engineers play a crucial role in improving living standards throughout the world (WFEO 2002). As a result, engineers can have a significant impact on progress toward sustainable development.

To help the public better understand the roles of engineers, this summary and the information on the attached CD have three goals:

1. To describe the engineer's contributions to sustainable development.

2. To outline significant accomplishments engineers have made toward sustainability since the Rio Summit of 1992.

3. To summarize ways that engineers can more effectively meet the goals of sustainable development.

---

[2] This section is a summary of a unique report, prepared for the World Summit on Sustainable Development (WSSD) in Johannesburg, South Africa, September 2002, by the Committee on Technology (ComTech) of the World Federation of Engineering Organizations (WFEO). WFEO represents 15 million engineers in 89 nations. The primary purpose of WFEO's ComTech is to share information on sustainable technologies. The CD, sponsored by the National Academy of Sciences and the National Science Foundation, and developed by CH2M HILL, one of the world's largest engineering and environmental companies, is appended to this ASCE Report, includes the complete WFEO report. The reader is urged to use the CD, which contains valuable illustrative flow charts, photos, and references, to better understand the remarkable contributions of engineers in working toward sustainable development. This material is included in this report with permission.

The Roles of Engineers

Engineers are problem solvers who apply the world of scientist and empirical engineering experience to a great range of projects. Most of the projects meet basic human needs:

> Water supply
>
> Food production
>
> Housing and shelter
>
> Sanitation and waste management
>
> Energy development
>
> Transportation
>
> Communication
>
> Industrial processing
>
> Development of natural resources

Engineers also work to eliminate or reduce environmental problems:

> Cleaning up polluted waste sites
>
> Sitting and planning projects to reduce environmental and social impacts
>
> Restoring natural environments such as forests, lakes, streams, and wetlands
>
> Improving industrial processes to eliminate waste and reduce consumption
>
> Recommending the appropriate and innovative use of technology

Accomplishments in the Last 10 Years

Engineers through the world have been working hard to achieve the goals of the Rio Summit of 1992. Since the Summit, they have accomplished the following:

> Formed the World Engineering Partnership for Sustainable Development and environmental committees in many other engineering organizations

Developed environmental policies, codes of ethics, and sustainable development guidelines

Contributed to the Earth Charter

Partnered on joint programs with the World Bank, United Nations Environmental Program, and the Global Environment Facility

Helped develop breakthroughs in computer technology and communication networks

Started educational programs for engineering students and practicing engineers on applying sustainable development concepts in their work

Developed new approaches in industrial processes to reduce the use of resources in manufacturing and eliminate waste products

Future Goals and Objectives

Engineers believe that using existing knowledge, technology, and experience can solve many problems facing developing nations. In the next 10 years, engineers can be of even greater assistance in achieving the goals of sustainable development through:

Creating a comprehensive program to identify and provide the information that engineers in developing countries require to meet energy, water, food, health, and other basic human needs

Generating global education programs on sustainable development for students and practicing engineers, and encouraging engineers to become environmental generalists

Becoming actively engaged in the full range of decision-making processes in addition to performing projects

Improving methods for identifying and considering all of a project's environmental costs, impacts, and conditions throughout a project's life cycle

Creating programs to provide hands-on help, share knowledge, and provide assistance on technically viable, commercially feasible, and environmentally and socially sustainable projects in developing countries

Supporting well-crafted policies, creative application of engineering principles, and a commitment to partnerships with social and physical scientist and health and medical professionals

Johannesburg Summit 2002 – Commitments and Implementation Initiatives[3]

The principal agreements, commitments, and implementation initiatives resulting from the World Summit on Sustainable Development (WSSD), held in Johannesburg, South Africa, in September 2002, are listed in this section (WSSD 2002). This list is not exhaustive, but reflects some key highlights of the summit process. The commitments shown are those agreed to in the Implementation Plan adopted by governments at the close of the Summit. Many of these initiatives will require significant engineering contributions and leadership.

Water & Sanitation

Commitments

Commitment to halve the proportion of people without access to sanitation by 2015; this matches the goal of halving the proportion of people without access to safe drinking water by 2015.

Implementation Initiatives

The United States announced $970 million in investments over the next three years on water and sanitation projects.

The European Union announced the "Water for Life" initiative that seeks to engage partners to meet goals for water and sanitation, primarily in Africa and Central Asia. The Asia Development Bank provided a $5 million grant to UN Habitat and $500 million in fast-track credit for the Water for Asian Cities Programme.

The UN has received 21 other water and sanitation initiatives with at least $20 million in extra resources.

---

[3] This section contains some of the agreements reached, and the commitments and implementation initiatives announced, during the World Summit on Sustainable Development, held in Johannesburg, South Africa, in September 2002. All material from the Summit can be found at: http://www.johannesburgsummit.org/ This material is included in this report with permission.

Energy

Commitments

> Commitment to increase access to modern energy services increase energy efficiency, and increase the use of renewable energy.

> To phase out, where appropriate, energy subsidies.

> To support the New Partnership for Africa's Development (NEPAD) objective of ensuring access to energy for at least 35% of the African population within 20 years.

Implementation Initiatives

> The nine major electricity companies of the E7 signed a range of agreements with the UN to facilitate technical cooperation for sustainable energy projects in developing countries.

> The European Union announced a $700 million partnership initiative on energy and the United States announced that it would invest up to $43 million in 2003.

> The South African energy utility Eskom announced a partnership to extend modern energy services to neighboring countries.

> The UN has received 32 partnership submissions for energy projects with at least $26 million in resources.

Health

Commitments

> Commitment that by 2020, chemicals should be used and produced in ways that do not harm human health and the environment.

> To enhance cooperation to reduce air pollution.

> To improve developing countries' access to environmentally sound alternatives to ozone depleting chemicals by 2010.

Implementation Initiatives

> United States announced commitment to spend $2.3 billion through 2003 on health, some of which was earmarked earlier for the Global Fund.

The UN has received 16 partnership submissions for health projects with $3 million in resources.

Agriculture

Commitments

The Global Environment Facility (GEF) will consider inclusion of the convention to Combat Desertification as a focal area for funding.

In Africa, development of food security strategies by 2005.

Implementation Initiatives

The United States will invest $90 million in 2003 for sustainable agriculture programs.

The UN has received 17 partnership submissions with at least $2 million in additional resources.

Biodiversity and Ecosystem Management

Commitments

Commitment to reduce biodiversity loss by 2010.

Commitment to reverse the current trend in natural resource degradation.

Commitment to restore fisheries to their maximum sustainable yields by 2015.

Commitment to establish a representative network of marine protected areas by 2012.

Commitment to improve developing countries' access to environmentally-sound alternatives to ozone depleting chemicals by 2010.

Undertake initiatives by 2004 to implement the Global Programme of Action for the Protection of the Marine Environment from Land Based Sources of Pollution.

Implementation Initiatives

The UN has received 32 partnership initiatives with $100 million in resources.

The United States has announced $53 million for forests in 2002-2005.

Cross-Cutting Issues

Commitments

Recognition that opening up access to markets is a key to development for many countries.

Support the phase out of all forms of export subsidies.

Commitment to establish a 20-year framework of programmes on sustainable consumption and production.

Commitment to actively promote corporate responsibility and accountability.

Commitments to develop and strengthen a range of activities to improve natural disaster preparedness and response.

Implementation Initiatives

Agreement to the replenishment of the Global Environment Facility, with a total of $3 billion ($2.92 billion announced pre-Summit and $80 million added by EU in Johannesburg).

Beyond Johannesburg – the US Engineering Community's Response[4]

A declaration by the US Engineering Community to the WSSD states:

"The survival of our planet and its people requires the collaboration of all professions in both developed and developing countries to sustain future generations. The goal of improving the social and economic well being of all people in the

---

[4] This section summarizes the activities to-date of the U.S. engineering community in responding to the challenges of Johannesburg, and developing a new role in sustainable development. A key response was *A Declaration by the US Engineering Community to the World Summit on Sustainable Development*," produced on June 24, 2002, at a meeting organized by the National Academy of Engineering, State Department, American Association of Engineering Societies, and American Institute of Chemical Engineers, in affiliation with the American Society of Civil Engineers, Engineers International Round Table, and World Federation of Engineering Organizations Committee on Technology. This material is included in this report with permission.

developed and lesser-developed countries is a pre-requisite for creating a stable, sustainable world. Although achieving this goal will require a broad coalition of well-crafted policies, it will only be realized through the application of engineering principles and a commitment to public/private partnerships involving professionals from all fields including the social sciences, engineering and medicine. It will also require collaboration for development, acceptance and dissemination of innovative solutions and better use of existing technologies.

Creating a sustainable world that provides a safe, secure, healthy life for all peoples is a priority for the US engineering community. It is evident that US engineering must increase its focus on sharing and dissemination of information, knowledge and technology that provides access to materials, energy, water, food and public health while addressing basic human needs. Engineers must deliver solutions that are technically viable, commercially feasible, and environmentally and socially sustainable.

Today's world is increasingly complex. The technical challenges required for sustainability have become enormous. The US engineering community must become engaged earlier in the policy formulation and decision-making process through its technical and profession societies to provide knowledge of environmental impacts, costs and feasibility. Engineers must be actively engaged in the entire decision making process from conceptualization, to project design, development and implementation. This includes the interdisciplinary process of building the evaluation/decision framework and the institutional infrastructure to realize a sustainable future.

We, the undersigned, commit to moving forward in support of the US engineering community to meet societal needs through capacity building, improved education, training, information development and dissemination, and engaging the engineering profession in all stages of the decision process.

Agreed, on the occasion of the Engineers Dialogue on Johannesburg and Beyond, June 24, 2002. The undersigned are participants in the dialogue and do not necessarily reflect the opinions of their stated associations." (US Engineering Community 2002)

This statement was endorsed by:

American Association of Engineering Societies

American Institute of chemical Engineers

ASME International – Environmental Engineering Division

National Academy of Engineering

National Society of Professional Engineers

Following this declaration, representatives from the World Bank, the Department of Energy, USAID, the National Research Council, the Environmental Protection Agency, and the National Academy of Engineering convened at an all day meeting on November 1, 2002 at the National Science Foundation. They discussed current developments and issues in their areas or responsibility, including water and sanitation, energy, health, and ecosystem management. Crosscutting issues such as engineering education, communications, disaster reduction, and mega-cities, were also discussed. Each panelist identified the important and essential roles that engineers could play in addressing current sustainability needs worldwide.

The US Department of State identified five partnership initiatives at the Johannesburg Summit. They are:

> "*Water for the Poor Initiative*" to expand access to clean water and sanitation services, improve watershed management, and increase the efficiency of water in industrial and agricultural activities (US to invest $970 million over three years);

> "*Initiative to Cut Hunger in Africa*" to spur technology sharing for small landholders, strengthen agricultural policy development, fund higher education, support regional technology collaboration, and expand resources for local infrastructure in transportation, marketing and communications (US to invest $90 million in 2003);

> "*Congo Basin Forest Partnership*" to promote economic development, alleviate poverty, improve governance, and conserve natural resources in six Central African countries (US to invest up to $53 million over the next four years);

> A health initiative to help fight HIV/AIDS, tuberculosis and malaria through financial and technical support (US Administration has requested $12 billion in 2003)

> The US engineering community is developing a multi-disciplinary partnership network that can effectively serve global and national sustainability programs and projects, and work towards meeting the goals of it s declaration. The objective is to create successful sustainable development projects in developing countries by establishing a broad program of targeted technical assistance carried out by the US engineering community. This program will team US engineers with their counterparts in developing countries through sister academies as well as other means.

As a next step, representatives of the US engineering community committed to the declaration will work with those responsible for implementing these partnership initiatives as well as other sustainability programs. This collaborative effort will determine how engineers can best contribute its specialized knowledge and training to meeting the needs of a safe, healthy and sustainable world.

SIDEBAR

Policy on the Role of the Engineer in Sustainable Development, American Society of Civil Engineers (ASCE 2001)

Sustainable development is the challenge of meeting human needs for natural resources, industrial products, energy, food, transportation, shelter, and waste management while conserving and protecting environmental quality and the natural resource base essential for future development.

The code of Ethics of the American Society of Civil Engineers (ASCE) requires civil engineers to strive to comply with the principles of sustainable development in the performance of their professional duties. Further the ASCE 2001 Action Plan calls for the Society to "promote sustainable development in project planning and development activities."

Therefore, ASCE supports sustainable development and will work on a global scale to promote public recognition and understanding of the need for sustainable development. ASCE will work to develop and encourage use of technologies required to achieve a sustainable world for future generations. ASCE supports the following implementation strategies.

Engineers should:

Cultivate a broader understand of political, economic, technical, and social issues and processes related to sustainable development;

Acquire the skills, knowledge, and information to facilitate a sustainable future;

Develop the tools required to achieve sustainable integration of the environment and development, together with other scientists and practitioners;

Develop economic approaches that recognize natural resources and our environment as capital assets;

Move beyond their disciplines to evaluate alternatives and to effect policy changes toward sustainable development;

Develop project teams with other design professionals, economists, and social, environmental, and physical scientists to arrive at sustainable solutions;

Adopt and apply an integrated systems approach for project decisions in which costs, benefits and effects on sustainability are considered for the whole lifetime and enduring effects of the project.

Work cooperatively with other trade and professional organizations that are focused on this issue to minimize duplication and bring the greatest resources to bear on advancing sustainability;

Utilize life costing in projects, which includes the associated environmental costs.

Issue

Economic and social development needs of both developed and developing countries are placing demands on natural resources that are fast outstripping supply. Furthermore, the ability of natural systems to assimilate wastes is being taxed, almost to exhaustion. Consequently, environmental, economic, social, and technological development must be seen as interdependent activities in which industrial competitiveness and sustainability are addressed together as complementary aspects of a common goal. Often, the impoverished bear disproportionate costs, and as a consequence, significantly impede progress toward sustainability.

The nature of development in the future will demand a new role for engineers and finding new ways to do business (i.e., doing more with less—less resources, less energy consumption, and less waste generation). It requires focus on upstream prevention in preference to "end of pipe" treatment, new manufacturing processes and equipment, expanded use of recyclable materials, and the development of regenerative/recyclable products and packaging. Sustainable development requires approaches that imitate natural or biological processes.

Rationale

Humanity needs a sustainable future. Engineers have a significant role in planning, designing, building and maintaining it. Engineers provide the bridge between science and society. In this role, engineers must participate in interdisciplinary teams with owners, suppliers, investors, regulators, community interest, ecologists, sociologists, and professionals from other disciplines, in achieving sustainable strategies and solutions.

Chapter II
Selected Readings on Sustainable Engineering Practice

This Chapter provides a broad spectrum of examples, which describe how sustainability principles are being integrated and applied in engineering education, research, and practice.

Sustainability in Engineering Education[5]

This reading is included to acquaint the reader with the importance of integrating sustainability into engineering education, and the issues and challenges involved.

Introduction

Sustainability is included in the new ABET Engineering Criteria as part of Criterion 4 which states in part "The curriculum must prepare students for engineering practice culminating in a major design experience based on the knowledge and skills acquired in earlier coursework and incorporating engineering standards and realistic constraints that include most of the following considerations: economic, environmental, sustainability, manufacturability, ethical, health and safety, social, and political." (Underlining added for emphasis) (ABET 2003)

Comments on the new Criteria and on sustainable development by engineers and others underline the lack of agreement on the meaning of sustainable development and sustainability in engineering. What sustainability means in engineering practice continuous to be debated, but students will benefit by joining in the continuing discussion of the issues surrounding sustainable development. Sustainable development in any form will not succeed without well-informed input from the engineering community. Although some of the environmental problems that have turned our attention to sustainable development are undoubtedly technology induced, effectively dealing with these issues and indeed fully implementing sustainability in practice will depend on new technologies and new technological approaches. The report by the National Academy (NRC 1999) notes that relatively little of the input over the last 10 years or so had been from the science and ~~technology~~ community.
engineering

---

[5] This section was prepared by William E. Kelly, P.E., Fellow, ASCE; Professor of Civil Engineering: The Catholic University of America; and Past President, Engineering Accreditation Commission, Accreditation Board for Engineering and Technology (ABET). This material is included in this report with permission.

The ASEE Board of Direction approved the following statement on sustainable development education in June 1999:

"Engineering students should learn about sustainable development and sustainability in the general education component of the curriculum as they are preparing for the major design experience. For example, studies of economics and ethics are necessary to understand the need to use sustainable engineering techniques, including improved clean technologies. In teaching sustainable design, faculty should ask their students to consider the impacts of design upon U.S. society, and upon other nations and cultures. Engineering faculty should use systems approaches, including interdisciplinary teams, to teach pollution prevention techniques, life cycle analysis, industry ecology and other sustainable engineering concepts.

Case studies, including studies of university-industry-government partnerships, can be used to illustrate the importance of the multidisciplinary aspects of designed systems, the impacts of those systems upon society and the environment, and the practical viability of the sustainability concept.

ASEE believes that engineering graduates must be prepared by their education to use sustainable engineering techniques in the practice of their profession and to take leadership roles in facilitating sustainable development in their communities." (ASEE 2002)

Clearly, there is a great deal of support and encouragement to move ahead with sustainability in engineering education.

Sustainability in General Education

As noted in the ASEE Statement, much of what students learn about sustainable development and sustainability must come from the general education component of the curriculum as they prepare for the major design experience. Studies of economics and ethics are necessary to an understanding of sustainable development and where technology, in particular clean technologies, could fit in (Goodstein 1995).

A number of the ABET outcomes (Criterion 3) fit very well with building a knowledge of sustainability as a consideration in design (ABET 2003). Overall, the ABET Criteria provide a framework for including sustainability in an undergraduate engineering program and suggest opportunities to educate the rest of the campus on sustainability issues as well.

For those interested in potential linkages between security and sustainable development, Kaplan (2000) and others have suggested that the environment could be the national security issue of the twenty-first century. The reasoning is that some of the parts of the world where water is in short supply are also some of the most unstable.

Sustainability in Engineering Design

The only explicit requirement for sustainability in the ABET Criteria is in the general criteria. With the great strides being made in the application of sustainability concepts to manufacturing processes, it is important to begin to define the principles that apply to all fields of engineering and some basic principles that could be introduced to all engineering students early in their engineering program.

In discussing general principles, it is important to define the design paradigm so as to include the considerations called for in ABET Criterion 4. It will also be necessary to define the systems context for design and define metrics that can be used to determine whether design goals are met and to compare alternatives.

Life cycle design is one of the ways the engineering profession could contribute to sustainable development. Although life-cycle design is far from normal practice, consideration of life-cycle costs is common in many areas. Pollution prevention is an element in life-cycle design and many companies are finding that pollution prevention just makes good business sense.

In teaching design, sustainability is a broad enough concept to encourage students to consider design in its societal and global context. This should also encourage students to think in more interdisciplinary ways. In teaching pollution prevention for industry, it is necessary to know industrial processes in some detail. In discussing pollution preventions in manufacturing it becomes very clear to civil engineering students that they are dealing with an interdisciplinary problem that requires a team effort for solution.

Consideration of systems aspects of sustainability (Roberts 1994) in engineering design education could encourage a reconsideration of the systems approach which has never caught on in engineering education in the United States. In this respect it is interesting to note that Bordogna (1998) has called for a new breed of civil "systems" engineers.

In discussing sustainability as a consideration in design, students must consider the context of their other projects. During construction of America's infrastructure, from the construction of the transcontinental railroad to the interstate highway system, relatively little consideration was given to the impacts that these systems would have on the environment and social-economic systems. Today, some even say that if we had had to consider sustainability when building these projects, they would never have been built!

By including sustainability in discussions of design, there is ample motivation to review some of the large existing infrastructure projects from the perspective of sustainability. It should be possible to learn a great deal from such an "enhanced" case history approach and at the same time learn about the history of their profession.

After World War II, Europe and Japan has the opportunity to rebuild much of their infrastructure. To what extent did this rebuilding allow them to build a more efficient infrastructure? What can we learn from this experience and apply to developing countries if sufficient capital can be found. Even with the best of intentions, sustainable development for developing countries is not going to be possible without capital. But efficient use of capital by well-educated engineers is going to be one of the keys to success in this development.

While ASCE includes some direction in its Code of Ethics, good examples of the application of the principles of sustainable development in civil engineering design, manufacturing, and construction are needed.

Academic institutions can contribute to the acceptance of sustainability in engineering design by presenting sustainability along with well-established considerations like economics, ethics, and the environment. Engineering departments have an opportunity, perhaps a responsibility, to take the lead in campus discussion of sustainable development. Whether sustainability will become the overarching issue that protection of the public health and safety is in design remains to be seen. However, it is a unifying concept that is not adequately addressed by environment, economics, and ethics considered separately.

Sustainable (Green) Buildings[6]

This reading presents a comprehensive overview of the principles, practice, and technology of planning, designing and constructing sustainable buildings.

Introduction

Sustainable buildings, also knows as Green Buildings, are important to sustainability in general and to the practice of engineering. Buildings provide shelter and support for most human activities. Their functionality, aesthetics, and healthfulness, safety, environmental quality and economy are vital to quality of life and productivity. In the United States (U.S. Bureau of the Census 2002), new construction and renovation of buildings together amount to about 1/11 of the Gross Domestic Product; the value of exiting buildings comprises about 48% of the

---

[6] This section was written by Richard N. Wright, with the advice and assistance of Barbara Lippiatt, an Economist with the National Institute of Standards and Technology (NIST). This material is included in this report with permission.

Nation's fixed, reproducible, tangible wealth; buildings consume about 40% of the Nations' energy and construction wastes are estimated to be 20-30 percent of the contents of landfills. Achievement of sustainability over the whole life cycle of a building (planning, design, construction, commissioning, operation, maintenance, renovation and removal) requires engineering expertise and provides an important market for engineering services and building products.

This reading provides qualitative understanding of the parameters affecting the sustainability of buildings and the techniques available to achieve sustainability, and gives references to sources of more detailed information. Sections are provided on environmental impacts, the building life cycle, environmental life cycle assessment, and sustainable building practices. It is based substantially on pioneering work in quantifying sustainability (Lippiatt 2002) for buildings performed by economist Barbara Lippiatt of the National Institute of Standards and Technology. Ms Lippiatt also generously advised in the preparation and review of this material.

Environmental Impacts

The Society of Environmental Toxicology and Chemistry (SETAC) has developed a general classification/characterization approach to environmental impact assessment (Owens et al. 1997), and the U.S. Environmental Protection Agency (EPA 2002) has implemented this approach. Impacts include: Natural Resource Depletion, Global Warming Potential, Ozone Depletion, Acidification Potential, Eutrophication Potential, Smog, Solid Waste, Erosion and Sedimentation, Ecological Toxicity, Human Toxicity, and Indoor Air Quality. Note many flows of resources in the building life cycle have multiple environmental impacts. For instance, use of electrical energy, generated from burning coal, for space heating and cooling contributes directly to all impacts except indoor air quality (and that is affected by the ventilation and energy conservation measures associated with space heating and cooling).

Natural Resource Depletion is central to sustainability. Natural resources consumed by a building include the land and habitat for endangered species of the building site, the building materials, nonrenewable energy consumed in providing the building materials and in space heating and cooling, water use, neighbors' access to natural lighting and views blocked by the building, etc. Land, for instance, is made unavailable for other uses, such as agriculture, during the life of the building, but may be recycled for other uses when the building is removed. Note that recycling has strong potential to mitigate natural resource depletion.

Global Warming Potential arises from the abilities of "greenhouse gases" in the atmosphere to reradiate energy to the earth's surface rather than allowing it to be lost to space. Greenhouse gases include water vapor, carbon dioxide, methane, chlorofluorocarbons, and ozone. These are produced in great volumes in the life cycles of buildings, with energy use for space heating and cooling generally the greatest contributor.

Ozone Depletion refers to thinning of the ozone layer in the stratosphere, which allows more harmful ultraviolet radiation to reach the earth's surface. Freon for refrigerators, heat pumps and air conditioners and halons for fire suppressants have been principal contributors to ozone depletion, but these are being replaced by less harmful refrigerants and fire suppressants.

Acidification Potential relates to harm to vegetation, soils, buildings, and people from acid rain or direct deposition of acidic particles. Buildings contribute to acidification principally from gases and particles released in energy generation for production, transportation and placement of building materials and products, and for space heating and cooling.

Eutrophication Potential arises from the addition of minerals and nutrients to soil or water, which lead to unnatural growth of organisms, such as algae, and loss of species of plants and animals. Buildings can contribute to eutrophication through discharges of nutrients in the manufacture of building products, through discharges of liquid and solid wastes from building services, and excessive fertilization of landscaping and building sites.

Smog arises from reactions of air emissions (from combustion, chemical reactions, out-gassing from materials, etc.) with sunlight to produce ozone and other light absorbing and irritating chemicals. Buildings can contribute to smog from: air emissions at the building (such as furnace exhaust gases and volatile organic compounds from building products), power plants providing energy to the building, occupants or customers traveling to the building, etc.

Solid Waste relates to the landfill volume of non-recycled wastes. Buildings contribute solid wastes in waste from manufacture of building products, wastes produced by building occupants, and in disposal of building materials and components during construction, maintenance, renovation and eventual removal of the building. Recycling, where possible, mitigates solid waste volumes (as when building trash is burned for power generation). Waste impacts other than volume are considered in eutrophication potential and in ecological and human toxicity.

Erosion and Sedimentation arise from storm water runoff, damage land and foul streams, rivers, lakes and oceans. Construction disturbs vegetation that absorbs moisture and restrains erosion, and buildings provide impervious ground cover that increases the amount of runoff.

Ecological Toxicity arises from detrimental effects of human activities on plants and animals beyond those encompassed in global warming, ozone depletion, acidification, eutrophication and smog. Buildings can contribute to ecological toxicity through the gaseous, liquid and solid emissions and wastes generated in: the manufacture, transportation and placement of building materials and components, the provision of building services, and activities of building occupants.

Human Toxicity is like ecological toxicity, but focused on detrimental effects on people. Buildings can contribute to human toxicity, for their occupants and near and distant neighbors, through the gaseous, liquid and solid emissions and wastes generated in: the manufacture, transportation, and placement of building materials and components, the provision of building services, and activities of building occupants.

Indoor Air Quality relates to the effects of contaminants in building air to the comfort, productivity and health of occupants. Contaminants can arise from emissions from building materials and equipment (such as space heaters and copying machines), cleaning and maintenance activities, emissions from the occupants themselves, and contaminants introduced by the ventilation system.

A great challenge to engineering for sustainability is that the above impacts, and others that may be important in a particular instance, are largely incommensurate. That is, there is no known and rational quantitative method to sum the impacts for a particular situation to arrive at an overall impact measure. BEES (Lippiatt 2002) provides a rational and systematic method to apply individual or group judgment to overall impact assessment. However, this situation is not unique to sustainability. For instance, for structural safety, incommensurate impacts of potential life loss, injuries, property damages and costs of safety measures, are weighed by experts and stakeholders in setting structural standards and codes. Similar deliberations can be used to develop standard practices for sustainable buildings.

The Building Life Cycle

Planning denotes the earliest phase of the building life cycle when its owners and designers determine the characteristics of the building and its site. These decisions establish the potential for sustainability of the building. For instance, the site will determine the transportation needs for the occupants and users of the building and for delivery of the building materials and components. Functional, aesthetic, health, safety, environmental and economic targets, which are called the "program," are set for the building. Generally, it is very difficult and expensive to make later changes to exceed these targets; these early decisions are crucial to sustainability and other qualities for the building.

Design denotes the decisions as to, and descriptions of, precisely what will be built. In conceptual design the appearance and layout are chosen, and decisions made on the types of materials and systems to be used. For instance, BEES (Lippiatt 2002) guides in selection of materials for sustainability. In preliminary design initial proportions and dimensions are identified as bases for detailed modeling of the building systems performance. In detailed design the final plans and specifications are prepared for the materials, components and systems.

The direct costs and environmental impacts of the planning and design phases are very modest, but the decisions made in planning and design have profound effects on the overall costs and environmental impacts of the building. For a typical office building with a 40 year service life (Hattis and Ware 1971) about 2% of the total life cycle cost provides the constructed building (planning and design typically are less than 10% of this 2%), operation and maintenance are 6%, and the salaries and overheads of the occupants are 92%. Land costs are not counted since the land remains valuable at the end of the life. Discounted to present value at 6% interest, initial design and construction is 5% of total life cycle cost, operation and maintenance is 6%, and occupant salaries and overheads are 89%. Taking environmental impacts as roughly proportional to costs, planning and design cause about 0.5% and set the potential for the remaining 99.5%. The potential for sustainability is dominated by decisions made in planning and design.

Construction denotes the preparation of the site, and purchase, transportation, and placement of the building materials, components and systems. Sustainability should be considered in the planning and management of construction – for economy, safety of the workforce, compliance with environmental regulations, and for the sake of sustainability itself. A major contribution to sustainability comes from conformance to plans and specifications. For instance missing or poorly installed insulation or gaps in heating and cooling ducts can greatly increase energy use for space heating and cooling with resulting detrimental environmental impacts and increased costs of operation. Also, change orders may make substitutions of less green materials.

Commissioning denotes the collaboration of designers, contractors, and building owners and operators to make the building operate as it was intended and designed to operate. Training in intended operation procedures, and tuning of the building's systems and controls are normal aspects of commissioning needed to achieve functional, safe and environmentally appropriate operations. Indeed, creation of an operating and maintenance manual is as appropriate in building design as it is in automotive design.

Operation and Maintenance denote the routine operations of the building during its service life. Effective operation and maintenance are vital to sustainability and other purposes and of the building. These deserve engineering attention in design and during the whole of the service life.

Renovation denotes major changes in or renewal of building systems to deal with replacement of major components, changes in the use of the building, or both.

Renovation involves stages of planning, design, construction and commissioning similar to those for the new building.

Removal denotes the deconstruction of the building and the return of the site to nature or its preparation for another use. Too often, removal occurs by abandonment of the vacant and derelict building with more severe environmental and social impacts than would result from a well-engineered process of removal.

Environmental Life Cycle Assessment

Environmental life cycle assessment (LCA) is an evolving, internationally standardized (ISO 14040 1997; ISO 14041 1998; ISO 14042 2000) method for sustainability assessment. All stages of the life cycle of a product must be considered for generation of environmental impacts. A product can be as simple as a nail or as complex as a subway system. Life stages for products include: raw or recycled materials acquisition, product manufacture, transportation, installation, operation and maintenance, recycling and waste management. All significant environmental impacts should be considered. The LCA methodology involves four steps: (1) goal and scope definition for the study, (2) inventory analysis of environmental inputs and outputs over the life cycle, (3) impact assessment for the inputs and outputs (carbon dioxide emission is an output, but its global warming effect is an impact), and (4) interpretation to combine the impacts (it is very challenging to combine different impacts for rational assessment).

BEES (Lippiatt 2002) provides a thorough and systematic method for the relative sustainability and economic assessment of 200 building products using Windows-based decision support software. It implements the LCA methodology of the ISO14040 standards and the ASTM E917 (ASTM 1994) life cycle cost method using environmental and economic performance data. Environmental impacts are combined with economic facts using the ASTM E 1765 method for Multi Attribute Decision Analysis (ASTM 1995). However, the combinations of environmental impacts are based in part on the judgment of some group of "experts" as is the relative weighting of environmental and economic facts. ATHENA Sustainable Materials Institute (ATHENA 2004) offers a similar approach, but it does not provide for combinations of environmental impacts with economic performance.

The life-cycle methods are intended to be used in building design to aid in the environmentally and economically responsible selection of building materials – such as the choice of roofing or floor covering. Their rigor is feasible because the factors affecting the decision can be well defined in advance and in general. Manufacturers have cooperated in development of the needed product data because many potential customers for product selection can use them. Note that these assessments cannot be directly applied outside of the United States because environmental regulations and manufacturing and transportation practices are different. However, similar life cycle methods have been used in other countries.

The Life Cycle Assessment and Multi Attribute Decision Analysis techniques can be used by standards organizations in preparing guidelines and standards for other decisions for sustainability such as: siting of buildings, selection of heating, ventilation and air-conditioning systems, or selection of structural systems. In the interim for engineering decisions for sustainability for individual buildings, a variety of less quantitative guidelines are available and are discussed in the next section.

Summary on Sustainability Assessment

1. International standards exist for environmental life cycle assessment and should be   followed, or exceptions taken noted explicitly.

2. All stages of the life cycle of the product or project should be considered.

3. All significant environmental impacts should be considered.

4. Judgment is involved, even in the most rigorous assessments. The engineer should determine that the judgments, implicit or explicit, in any assessment method used are appropriate to the product or project assessed.

Sustainable Building Practices

Many organizations have prepared guidance and recommendations for sustainable building practices. The recent guide prepared by the U.S. Federal agencies collaborating through the National Academies' Federal Facilities Council (FFC 2001), in addition to its own guidance, lists eleven pages of references to "on-line" resources for sustainable building practices. Among leading resources are the U.S. Green Buildings Council's LEED Green Building Rating System (USGBC 2004), the Environmental Building News' priorities for green building (EBN 2004), and the Whole Building Design Guide (WBDG) of the National Institute of Building Sciences (NIBS) (WBDG 2004).

LEED and WBDG priorities are similar. Quoting the WBDG:

> **Optimize Site Potential.** Creating sustainable buildings starts with proper site selection, including the reuse or rehabilitation of existing buildings. The location, orientation, and landscaping of a building affect the local ecosystems, transportation methods, and energy use.

> **Minimize Energy Consumption.** A building should rely on conservation and passive design measures rather than fossil fuels for its operation. It should meet or exceed applicable energy performance standards.

Protect and Conserve Water. In many parts of the country, fresh water is an increasingly scarce resource. A sustainable building should reduce, control or treat site runoff, use water efficiently, and reuse or recycle water for on site use when feasible.

Use Environmentally Preferable Products. A sustainable building should be constructed of materials that minimize lifecycle environmental impacts such as global warming, resource depletion, and human toxicity. In a materials context, life cycle includes raw materials acquisition, product manufacturing, packaging, transportation, installation, use and reuse/recycling/disposal.

Enhance Indoor Environmental Quality. The indoor environmental quality (IEQ) of a building has a significant impact on occupant health, comfort and productivity. Among other attributes, a sustainable building should maximize daylight, have appropriate ventilation and moisture control and avoid the use of materials with high VOC emissions.

Optimize Operational and Maintenance Practices. A sustainable building should be designed to take into account the energy and environmental impacts of operating and maintaining the building. Designers are encouraged to specify materials and systems that reduce the need for maintenance, and/or require less water, energy, and toxic chemicals and cleaners to maintain.

Environmental Building News lists additional priorities:

Create Community. Design communities to reduce dependency on the automobile and to foster a sense of community.

Reduce Material Use. Optimize design to make use of smaller spaces and utilize materials efficiently.

Maximize Longevity. Design for durability and adaptability.

Minimize Construction and Demolition Waste. Return, reuse and recycle job-site waste.

Green Up Your Business. Minimize the environmental impact of our own business practices.

Sustainability Applications in a Federal Agency[7]

This reading provides an example of how the principles of sustainability are being applied in a mission-oriented Federal Agency, the U.S. Army Crops of Engineers (USACE).   Included are a set of USACE Environmental Operating Principles (USACE 2004); a long-range sustainable planning exercise at Fort Bragg (Ft. Bragg 2004); a modification of the U.S. Green Building Council's LEED's rating system, the Sustainability Project Rating Tool (SPiRiT 2004); and a "green" facility study at Fort Hood, Texas (Doyle and Salmon 2004).

USACE Environmental Operating Principles

1. Strive to achieve environmental sustainability. An environment maintained in a healthy, diverse and sustainable condition is necessary to support life.

2. Recognize the interdependence of life and the physical environment. Proactively consider environmental consequences of Corps programs and act accordingly in all appropriate circumstances.

3. Seek balance and synergy among human development activities and natural systems by designing economic and environmental solutions that support and reinforce one another.

4. Continue to accept corporate responsibility and accountability under the law for activities and decisions under our control that impact human health and welfare and the continued viability of natural systems.

5. Seek ways and means to assess and mitigate cumulative impacts to the environment; bring systems approaches to the full life cycle of our processes and work.

6. Build and share an integrated scientific, economic, and social knowledge base that supports a greater understanding of the environment and impacts of our work.

7. Respect the views of individuals and groups interested in Corps activities, listen to them actively, and learn from their perspective in the search to find innovative win-win solutions to the nation's problems that also protect and enhance the environment.

---

[7]   This section was written by Albert A. Grant. This material is included in this report with permission.

Long Range Sustainability Planning at Fort Bragg

Fort Bragg held an Environmental Sustainability Executive Conference in April 2001. The purpose was to find ways to meet the mission of combat readiness while maintaining environmental stewardship. Participants included regulators and members of the local community.

The ten goals established at the April conference will form the basis for the Integrated Strategic Environmental Plane (ISEP), the blue print for the installation's long-term success in terms of infrastructure and natural resources.

Participants at the conference were asked to develop goals "based on an integrated approach that includes all stakeholders – military, civilian, communities, and regulatory agencies – to achieve a meaningful consensus and ensure readiness."

Given these challenges, community members identified the following goals for Fort Bragg:

> Challenge: Use of energy at Fort Bragg, whether it's generated on post or off, contributes to the high levels of ozone in the air. Further, the events this winter in California and across the nation raise serious concerns about the cost of energy and the availability of energy at any cost. How can fort Bragg protect and secure the energy it needs to operate?

> o   Goal 1: Eliminate energy waste by giving commanders energy goals and data on actual energy use by 2002.

> Challenge: Facility construction, operation, maintenance, and demolition are costly, leading to numerous environmental impacts and large energy and water use. How can Fort Bragg provide the world-class facilities that soldiers and families deserve, while minimizing associated pollution, resource depletion, and costs?

> o   Goal 2: Design all new construction to Leadership in Energy and Environmental Design (LEED) platinum standard by 2006.

> Challenge: The state of North Carolina is increasingly concerned about ozone and other air pollutants. How can Fort Bragg minimize future costs and operational restrictions while improving regional air quality?

> o   Goal 3: Develop acceptable regional commuting options by 2025.

> o   Goal 4: Operate 100% of non-tactical fleet on alternative fuels by 2010.

Challenge: Potential sources of water for Fort Bragg consumption have been steadily declining (both in quantity and quality) due to overuse. How can Fort Bragg reduce its dependence on these sources and provide premium quality drinking water, as well as the "right" quality water for other uses, without aggravating future regional water supply issues?

- Goal 5: Reduce water consumption 90% by 2025.

Challenge: Contamination of regional water resources, particularly by sediments, is a critical consideration to North Carolina, because of the economic impacts associated with destruction of fish habitats, treatment of water to drinking quality, and the decrease of drinking water reservoir holding capacity. How can Fort Bragg minimize the future costs and potential operational restrictions associated with water pollution, while improving regional water quality?

- Goal 6: Ensure water quality leaving Fort Bragg is equal to or better than water quality coming onto post by 2025.

Challenge: Fort Bragg buys $176M worth of products and materials every year – and throws away over 200,000 tons at a total cost well over $3M. How can Fort Bragg promote the sustainable manufacture, use and disposal of materials and products, while minimizing costs and environmental impact? How can Fort Bragg stimulate local and national markets for environmentally preferred products?

- Goal 7: Landfill zero waste by 2025.

- Goal 8: Buy 80% environmental preferable products from local sources by 2025.

Challenge: Fort Bragg maintains 161,597 acres of land for training. Of this only 72,236 are unrestricted for use. How can Fort Bragg provide enough usable land for military training – and ensure that training is not further constrained by concerns over potential environmental contamination and negative impacts on endangered species? How can Fort Bragg use its land requirements to address the effects of urban sprawl and regional needs for open space and biodiversity?

- Goal 9: Educate 100% of personnel on environmental responsibilities, to cut enforcement actions to 0 by 2002.

- Goal 10: Adopt compatible land use laws/regulations with local communities by 2005.

After the July meetings, there was an opportunity for the public to get involved in the process. State and federal regulatory agencies, as well as local community officials, were encouraged to share ideas and build partnerships as Fort Bragg works toward sustainability planning.

SPiRiT

The Army has taken a cue from the private sector and developed a military version of the Green Building Council's LEEDS (Leadership in Energy and Environmental Design) model, knows as SPiRiT – Sustainability Project Rating Tool. This model, developed under the auspices of the USACE Engineer Research and Development Center (ERDC), Construction Engineering Research Laboratory (CERL), is now being used to evaluate our military construction projects in terms of their sustainability or, in other words, how well they incorporate "green" building techniques, such as recyclable building materials, energy efficiency, natural daylight and compatibility with the natural surrounding. The Assistant Chief of Staff for Installation Management and the USACE have decided to strive to achieve a Bronze rating for all projects, with several being singled out as showcase projects at the Gold or Platinum level. This initiative is a common sense design and building practice that is intended to reduce lifecycle costs while helping the Department support other federal goals, like energy reduction.

Fort Hood's First "Green" Facility

The Fort Hood Directorate of Public Works (DPW), in concert with Steinbomer and Associates Architects, Bragg Landscape, Fire Protection Engineering, Beneco Enterprises, Jaster-Quintanilla & Associates, Way consulting Engineers, HMG Engineering Associates, Austin Energy's Green Building Program and the Army Corps of Engineers, has partnered to design and build Fort Hood's first ever "green" facility. The Fort Hood Environmental Training Facility is scheduled to be the first of its kind to earn the COE's SPiRiT Platinum certification level. Platinum rating is the highest certification level achievable. This new 5000 square-foot multi-purpose training facility will be located in the southwest part of main Fort Hood, near the current DPW headquarters. This required a dedicated team effort that capitalized on sustainable methods and practices while integrating new energy management technologies and methods.

The facility will combine the latest in energy management technologies, while encompassing sustainable design concepts. Part of the floor is salvaged from a recently demolished bowling alley. The exterior walls will be made of straw bales, and the sand for the stucco will be ground bottles from Fort Hood's recycling center. The facility will utilize waterless urinals and low flow toilets to conserve water. In addition, rainwater collection will be used for a drip irrigation system. The landscaping design will utilize low maintenance, local vegetation while meeting

FORSCOM force protection requirement. For further energy management practices active day lighting will be used, along with motion sensors throughout the facility to turn lights off when not needed.

The orientation of the building is set to maximize the local weather patterns for cooling. The insulation factor of the straw bales, combined with the highly efficient windows will provide a highly efficient structure. They anticipate a reduced need for conditioned air during the long hot Texas summers.

This project has been a valuable educational experience for Fort Hood. They quickly realized that they couldn't do it all. Trade-off decisions were made based upon desired sustainability versus budgetary restrictions. Life cycle cost analysis was used to determine energy management methods that would give the biggest bang for the buck, while earning enough points to achieve SPiRiT certification. Another important element is patterning the project to the areas. Full length porches on the south, a breezeway to capture wind, double hung windows and a metal roof all help keep the building comfortable in the hot, humid climate of Central Texas. The use of low-emitting materials was patterned after local or state regulatory requirements. The success of the project was a direct result of an energized team that had experience with sustainable designs and projects, and was motivated to think "outside" the box.

Sustainability Concerns and Mitigating Strategies for Transportation[8]

This reading discusses transportation sustainability issues and concerns on a global scale, and mitigating strategies that have been developed, tested, and implemented, all within the broad context of transportation mobility needs.

The World Business Council for Sustainable Development is a coalition of 150 international companies united by a shared commitment to sustainable development via the three pillars of economic growth, environmental protection, and social equity. Its mission is to provide business leadership as a catalyst for change toward sustainable development, and to promote the role of eco-efficiency, innovation and corporate social responsibility.

---

[8]   This reading is abstracted from the Report "Mobility 2001, World Mobility at the End of the 21st Century and its Sustainability," from the Sustainable Mobility Project (SMP) Working Group of the World Business Council for Sustainable Development (WBCSD 2001). The full report can be found at:
http://www.sustainablemobility.org/publications/publications.asp
In addition, further information on the Sustainable Mobility Project can be found at
http://www.wbcsd.org/web/mobility.htm
This material is included in this report with permission.

Sustainability Concerns

Though the widespread ownership and use of private automobiles produce widespread benefits at many levels – increased personal mobility, regional development, and greater access to social and economic opportunities – they are also responsible for a number of negative developments that raise serious sustainability concerns. An automobile-dominated transport system leaves some members of the society behind because they are too poor to afford a car or physically incapable of driving. Urban motorways, built to accommodate increased traffic, divide and disrupt urban neighborhoods. The rising use of motor vehicles and rising travel demand can lead to deteriorating air quality and greater emissions of greenhouse gases, accelerated depletion of fossil-fuel reserves, higher accident rates, and mounting traffic congestion. Public opinion everywhere views traffic congestion as the consequence of motorization that most seriously affects the environment and the quality of life. Congestion and delay not only waste time and are a major irritant to individual travelers; they also burden businesses and the economy with higher costs. Indeed, the vitality of national economies is intimately linked to a smoothly functioning transportation system. The following summarizes the nature of those concerns at the beginning of the twenty-first century. It first reviews those aspects most closely tied to vehicle technology (accidents, emissions, noise, etc.), then looks to wider social impacts, including increased levels of traffic congestion.

Road Safety

The cost in human lives, injuries, and suffering attributable to highway and road crashes in enormous. Toward the end of the 1990's, around 42,000 people were killed each year in road accidents in Western Europe and between 40,000 and 45,000 in the United States. In some countries, road accidents are the primary cause of death in the 15- to 30-year-old age group. The number of people seriously injured in road accidents is typically more than ten times higher. Road accident victims are not just motorized vehicle drivers and occupants, but also include pedestrians and bicyclists. In developed countries, these groups account for roughly 10% to 15% of the total number of road fatalities. Between half and three-quarters of the highway fatalities in developed countries occur outside of urban areas; however, even one-quarter of these totals attributable to motor vehicle accidents in urban areas is a large amount.

There are various ways of estimating the monetary costs to society of road accidents; the different methods vary in the way that they account for pain and suffering and economic losses. Estimates of accident costs in most developed countries are typically in the range of 1% to 3% of GDP.

During the last decade, all industrialized countries have made important strides in reducing highway fatalities – down 25% in Western Europe and 30% in the United States – a trend attributable to a combination of safer vehicles (dual air bags, mandatory safety belts, antilock brakes, etc.), safer highway designs, faster incident response, and better post-accident care. All of these trends are expected to continue into the foreseeable future.

Nonrenewable Resource Consumption

Current vehicle propulsion technologies are based on the combustion of petroleum-based fossil fuels. Transport is not only the major, but also the most rapidly growing sector of oil consumption in countries within the Organization for Economic Cooperation and Development (OECD). Between 1973 and 1988, transport's share of total oil consumption on OECD countries grew from 43% to 60%. Light- and heavy-duty road vehicles – passenger cars, light trucks, motorcycles, and heavy trucks – use approximately 75% of all transport fuel. Through the effects of stricter government standards, higher fuel prices, and voluntary efforts by vehicle manufacturers, there have been notable improvements in vehicle fuel consumption and carbon dioxide emission rates over the past several decades. Much of the initial impetus for these efforts was provided by the oil crisis in the 1970's. The greatest initial reductions in fuel consumption predictably occurred in the countries with the most fuel-inefficient fleets, i.e., the United States, Canada, and Australia. In the United States, the fuel consumption of new light-duty vehicles (automobiles and light trucks) declined by an average of 3% per year between 1978 and 1987.

In the United States, average new light-duty vehicle fuel consumption reached its low point of 9 liters/100 kilometers in 1987. Since then, it has been increasing, in large part because of growing sales of light trucks, minivans, and sport utility vehicles (SUV's). While engine efficiency is continuously being improved, consumers often prefer to exchange some of all of the potential fuel efficiency gains for increased performance characteristics, such as increased power and torque. Moreover, changes in safety regulations sometimes lead to increased vehicle weight, resulting in higher levels of fuel consumption. By 1999, the average fleet new vehicle fuel economy had risen to 9.6 liters/100 kilometers.

Due to the slow rate of fleet turnover, the fuel consumption of the on-road fleet is higher than that of the new vehicle fleet whenever the latter is falling. In 1995 it is estimated that the fuel consumption of the on-road light-duty vehicle fleet was 9.5 liters/100 kilometers in OECD Europe, 11.4 liters/100 kilometers in OECD North America (the United States and Canada), and 10.0 1/100 kilometers in OECD Pacific (Australia, New Zealand, and Japan). Including the other road vehicle types (motorcycles and heavy trucks) raises threes figures to 10.1 liters /100 kilometers for OECD Europe, 13.5 liters/100 kilometers for OECD North America, and 13.3 liters/100 kilometers for OECD Pacific.

Carbon Dioxide Emissions

In the developed world, transport is one of the few industrial sectors for which CO_2 emissions are growing. It has been estimated that transport activities account for roughly 28% of total worldwide CO_2 production by humans. The share of transport in total CO_2 production is slightly higher in North America (roughly 33%) and slightly lower in Western Europe (roughly 24%). The United States alone accounts for roughly 24% of total global CO_2 emissions, so the US transport sector emits roughly 8% of the total world output of CO_2.

In 1998, the auto industry in Europe signed a Voluntary Agreement with the European Union to reduce corporate CO_2 emissions of the new car fleet from an average of 186 g/km in 1995 to 140 g/km in 2008. It is likely that manufacturers will rely heavily on sales of diesel engines and smaller cars to reach the target. However, total transport-related CO_2 emissions are closely connected to the total amount of fuel consumed. This depends, in turn, upon the number of transport vehicles, and the intensity of use of these vehicles in addition to the in-use efficiency of the average transport vehicle. Transport-related CO_2 emissions are expected to rise in all OECD countries at least through 2015. Average vehicle fuel consumption is expected to decline, but increases in the number of vehicles and in the intensity of their use are projected to overwhelm this decline.

Noxious Emissions

Concerns about local air pollution caused by automobiles were one of the original sustainability concerns related to automobile use in most of the developed world. During the last 30 years, most industrialized counties have adopted vehicle emission control measures in an effort to control and abate deteriorating air quality in urbanized areas. Preventive measures include exhaust emission standards, evaporative emission controls, fuel quality requirements, inspection and maintenance programs, and refueling controls. These control measure have resulted in substantial reductions in vehicle emissions, and in an overall improvement in air quality as measured by the declining number of days that pollutants in the air exceed recommended levels.

According to OECD estimates, HC and NO_X emissions in OECD countries will fall substantially during the period 2005-2010. Beyond that point, however, the OECD forecasts suggest if present trends continue emissions of HC and NO_X will start increasing as vehicle-kilometers traveled increase, and will be above their 2000 levels by 2030. For CO, if present trends continue, levels will fall until 2010, level off, and then start increasing again.

Vehicular Noise

Most developed countries have had vehicle noise emission regulations since the 1970s. Technological progress in engines and exhaust systems has made new vehicles considerably quieter. For example, the EU allowable noise level of a modern truck is approximately equivalent to that of the typical car in 1970. Nonetheless, according to a statement issued in June 1999 by European Health, Environment and Transport ministers, transport – and in particular road traffic – remains the main cause of human exposure to ambient noise. The proportion of the population in the European Region exposed to "high" noise levels (equivalent to 65 dBLAeq over 24 hours) increased from 15% to 26% between 1980 and 1990. About 65% of the European population is estimated to be exposed to noise levels leading to serious annoyance, speech interference, and sleep disturbance (55-65 dBLAeq over 24 hours).

Besides vehicle engines and exhausts, much of the noise produced by vehicles today, particularly in highway operation, results from the movement of vehicles through the air, and the contact of tires with the road. The former can be reduced by aerodynamic vehicle body designs (which also have the effect of improving fuel efficiency and reducing emissions). The latter can be reduced through tire tread designs and improvements in pavement surface textures (which also have the effect of draining water more effectively and so reducing accident risks). Noise barriers can also minimize the impact of vehicle-generated noise on nearby activities.

Economic Viability of Public Transport

The inability of local public transport across most of the developed world to recover costs raises a concern about its economic sustainability. Most local scheduled surface public transport operations in developed countries are unable to meet their operating costs from commercial revenues.

The total sums involved can be considerable. The United Kingdom, with perhaps one of the least subsidized public transport systems in the developed world, receives about US $3.5 billion per year in external funding, which is equivalent to US $0.50 per journey or US $0.04 per passenger-kilometer. Elsewhere, subsidy rates are usually significantly higher than this. In the United States, for example, governmental funding for public transport exceeds US $8billion per year giving support rates in excess of US $1.00 per journey and US $0.13 per passenger-kilometer.

In effect, public transport needs significant public subsidies to operate. There are certainly many reasons to provide public transport with public support, and there is little risk that the inability to recover operating costs would force the suspension of public transport service in any major metropolitan region in the developed world. However, subsidies and the resulting levels of service are often vulnerable to the vicissitudes of public opinion and varying levels of political support for public transport. Consider, for instance, the example of GO Transit, the public transport system serving the Toronto metropolitan area. Long heralded as a leading example of an effectively governed, well-managed system with high ridership in North America, GO (despite fairly high, mandated farebox recovery ratios) has been so starved of capital expenditures in recent year that is now faces serious service and maintenance problems.

Creation of Transport-Disadvantaged Social Groups

Because of the great reliance on private vehicles for transport in developed societies, people who cannot afford to buy or lease a vehicle, as well as those who, because of physical or mental handicaps, cannot operate one, may find themselves seriously disadvantaged in their ability to get to jobs or services, or to take care of other needs. Children and teenagers as well as elderly adults may also suffer from the automobile-orientation of the transport and land-use systems.

This effect is particularly perverse in the case of people kept by poverty from owning a private vehicle. The system limits their access to the very job opportunities that could provide the means of improving their economic situation and rising out of poverty.

Although many public transport systems are adapting vehicles and stations to accommodate the needs of riders with disabilities, the origins and destinations of these riders may not be conveniently served by the public transport system. Problems of access to, and egress from, the system are even more dissuasive for these riders than for the general public. Special-purpose, often door-to-door demand-responsive systems can meet some of their needs but generally require significant advance booking and do not provide tight schedule guarantees, so they may be less convenient in some ways that conventional public transport.

Increasing suburbanization and the living patterns of older adults suggest that the automobile will be of growing importance to the next generation of retirees. However, many older people have special transport needs that may not be well met by an automobile-dependent system. Countries in the European Union as well as Japan and the United States will experience significant growth of the population of older people in both real numbers and in the proportion to the total population. The number of people in Europe over 60 is expected to grow by 66% to 116 million in 2025, from a 1996 base of 77 million. In Germany, France, and the United Kingdom, between 15% and 17% of the population are over 60, with trends indicating rapid aging over the next two decades. Italy has the distinction of being the oldest nation within Europe – defined as the proportion of people over 60 coupled with the lowest fertility rate. Italy now has more people over age 60 than people 20 and younger. Similarly, Japan's population leads most of the world in the number of older adults. The number of Japanese over age 60 is expected to grow by 65% over the next 25 years, from 26 million to 40 million. The US baby-boom population totals nearly 80 million people. As this group ages, the number of people over 60 will nearly double in 25 years. There are more than 44 million people over age 60 today in the United States; their number is projected to increase to well over 80 million in 25 years.

As the population grows, cars driven by the elderly will constitute an increasing proportion of traffic, especially in the suburbs and in rural areas, where many elderly people tend to reside. Addressing the issue of million of drivers in their late 70s and beyond involves a tough tradeoff between safety and mobility, between the risk to life and the risk to quality of life. Overall, older drivers spend far less time on the road then younger drivers and have fewer accidents. On the basis of accidents per miles driver, however, they have a higher rate of crashes. The rate rises after age 75 and increases significantly after 85. The driver fatality rate shows that drivers over 75 years of age are more vulnerable than any other age group except teenagers. Factors like vision and hearing impairment, disorientation, memory disorders, and side effects from medicines are largely responsible.

Understandably, most elderly people want to hold onto their cars and driver's licenses as long as possible. A valid driver's license is as much a certificate of continued vitality and independence as it is a mobility necessity. To have to give up driving is viewed as a step toward dependency and isolation. While some countries have adopted stringent procedures aimed at identifying unsafe elderly drivers and getting them off the roads, public attitudes toward the treatment of elderly drivers are shifting, as the younger generation confronts the reality of sustaining the mobility of their own elderly parents. There is a growing sentiment that it is both unreasonable and unfair to expect elderly people to give up their cars, since doing so would be tantamount to sentencing them to a life of imprisonment in their homes. Instead, one increasingly hears demands that society must find ways to make it safe for older people to continue driving. To a significant extent, that is why the focus of debate has shifted in the last few years. While there still are calls to get unsafe older drivers off the road, far more attention is being paid to help elderly persons stay on the road.

Community Disruption

Although more difficult to quantify, the increasing orientation of the urban transport system toward private vehicles can have additional effects on the quality of community life. As we have seen, the more flagrant cases of this provided the initial rallying call for the "freeway revolt" against urban highway expansion in the United States. Urban motorways were sometimes built through the middle of established communities (most frequently those with insufficient political power to oppose that alignment), in effect dividing the community and constructing a physical barrier between the two halves.

More generally, in a community dominated by private-vehicle travel, there are relatively few opportunities for serendipitous interactions between residents, because when people leave their homes they isolate themselves in cars. This can lead to loss of sense of community and social cohesion. To attribute such developments entirely to automobility would be a gross distortion, yet there is a palpable, if inchoate sense that the increased use of cars over longer commutes has led to a more harried, less friendly society.

Traffic Congestion

The general public views traffic congestion as one of the most vexing drawbacks of a widely automotive society. Moreover, though highway traffic congestion is a subject that is sometimes sensationalized, there is growing evidence that congestion is increasing in intensity, duration, and geographic coverage across the developed world. According to the *2001 Urban Mobility Study* published by the Texas Transportation Institute, the average annual delay per person in the United States has climbed from 11 hours in 1982 to 36 hours in 1999. An OECD study that uses average daily vehicle-kilometers per road-kilometer as a proxy measure of congestion concludes that congestion is also increasing in Europe.

Further, there is evidence that congestion has gone from being essentially localized peaking problem – stemming from too many drivers trying to use a limited number of critical road links at critical times; i.e. "rush hours" – affecting a few drivers, to becoming a more widespread phenomenon that affects most drivers. For instance, a 1992 study of US congestion found that the percentage of peak-hour urban interstate travel that occurred in congested conditions increased from 40% in 1975 to close to 70% by 1990. Changing urban travel patterns, particularly the increasing role of suburb-suburb trips, are responsible for many of these trends. Congestion resulting from suburb-suburb trips has a different pattern from the localized severe "rush-hour" congestion on radial links to central business districts; it tends to spread in time and space.

The consequence of these trends is that urban residents are spending double and triple the amount of time in congested traffic that they did only 20 years ago. Wasting hours each week in congested traffic is a deeply frustrating and fruitless use of time for adult urban residents. The economic, social, and environmental costs to society of thousands of drivers driving slowly on congested highways and major thoroughfares are considerable; the drivers are unable to perform productive economic endeavors, or to engage in beneficial recreational activities with family and friends, while their automobiles produce substantial amounts of airborne pollutants. Unfortunately, because of the growth of suburbs and the inability of public transport to serve them, the staunch opposition to new highway construction, and individuals' visceral attachment to driving in their cars, an equitable, politically palatable solution to urban congestion remains elusive.

Mitigating Strategies

Every industrialized nation has worked to develop policies to mitigate the adverse effects of motorization without impairing the continued growth of mobility. In this section we describe the most common of these strategies; highlighting best practices and particularly successful applications. The mitigating strategies can be classified under six broad categories:

1. Reducing the demand for automobile use

2. Improvements – both physical and operational – in the provision of highway and public transport infrastructure

3. Improving the transport options available for travelers

4. Using innovative land-use and urban-design strategies to reduce travel demand

5. Integrated approaches that combine multiple strategies

6. "Civilizing" the motor vehicle by modifying its design to increase safety and crashworthiness and reduce emissions and fuel consumption.

Each of these broad categories includes multiple strategies. For instance, the demand for automobile use can be reduced in a number of ways: automobiles can be priced to reduce demand; more environmentally sound paradigms of vehicle ownership and use can be encouraged; and automobile use can be restricted. Similarly, supply-side improvements could include building new infrastructure, as well as operating and managing existing infrastructure more efficiently.

Reducing the Demand for Auto Use

Over the last three decades, the negative effects of the auto have spurred the creation of several strategies to ameliorate these effects by reducing the demand for automobile travel. They include direct restrictions on auto use as well as a number of innovative ideas that are more nuanced in their approach.

Transportation Demand Management

Transportation Demand Management (TDM) is a set of techniques that aim to reduce or redistribute travel demand, curtail solo driving, and decrease auto dependency. The typical TDM techniques include promotion of carpooling, flexible working arrangements, telecommuting, road pricing, and time-saving high-occupancy vehicle (HOV) lanes. In recent years, metropolitan regions in several developed countries have adopted TDM as part of their transportation plans. In the early 1970s, several US corporations initiated voluntary carpool programs in response to appeals to conserve fuel during the energy crisis. More recently, corporations have offered flexible working hours, telecommuting, and other actions aimed at reducing or redistributing commuter travel demand. California and certain metropolitan areas of the United States mandated ridesharing programs during the 1980s, but later terminated them in the face of widespread public opposition. Modest TDM programs also have been implemented in Australia, Canada, and the Netherlands.

Evidence over the past two decades reveals that demand management has a limited influence on travel behavior; not unusual are the results of an evaluation of efforts in the San Francisco Bay Area which estimated that "conventional" TDM (excluding road pricing options) had the potential to reduce total fuel use and emissions by about 7% to 8%.

City Center Automobile Restrictions

Automobile restrictions have won acceptance as a legitimate technique of congestion management and as an instrument of achieving sustainable mobility in crowded city centers. They are employed in more than 100 cities of Europe, North and South America, and Asia as documented by OECD surveys. Center-city restrictions very in duration, scope, and severity, ranging from temporary traffic prohibitions in commercial districts during shopping hours to permanent closure to vehicular traffic in entire historic town centers ("car-free zones"), as in Vienna, Austria; Munich and Bremen, Germany; and Bologna and Turin, Italy.

More drastic restrictions, involving outright bans on automobile use in center cities, have been instituted when pollution reaches unhealthful levels. In Athens, Mexico City, and during a September 1998 pollution inversion incident in Paris (cool air trapped by warm air above it, which keeps pollution from dispersing), authorities banned cars from entering the city center on alternate days, according to whether license plates ended with an odd or even number.

Traffic Calming

In residential areas, there are a variety of regulatory and physical "traffic-calming" measures used to slow down and discourage through-traffic. The roots of the movement to reduce or "calm" vehicular traffic can be traced to Western Europe, where concern about traffic and the political will to act upon it surfaced in the early 1970s. The Netherlands pioneered the concept of the *Woonerf* – protected residential areas in which pedestrians had absolute priority over vehicular traffic; and German cities introduced the concept of *Verkehrsberuhigung* (the origin and literal translation of the term "traffic calming") – a policy of limiting the use of autos in residential areas using an array of techniques, such as diverting through-traffic, limiting parking to designated areas, installing physical speed restraints and declaring certain areas off-limits to the automobile. In the last decade, many regions in the United States have adopted physical design measures to dissuade motorists from cutting through residential areas en route to and from work.

The Rebirth of the City Car

The launching of Renault's Twingo, Fiat's Cinquecento, Volkswagen's Lupo, and DaimlerChrystler's A-class and Smart models in the late 1990s marked a rebirth of the ultra-compact "city car" compatible with the crowded urban environment. Unlike their 1970s precursors that were spartan in appearance and aimed at young, budget-conscious drivers, the current generation of "minis" are designed to appeal to more affluent customers. Sales figures over the past several years, and the interest stimulated by the launching of DaimlerChrysler's Smart, suggest that the city car has found a market niche among European car buyers who want a small, maneuverable, and fuel-efficient car but also desire style, fun, and a good riving experience. This favorable market reception is being reinforced by government/industry-sponsored field tests of small electric and hybrid city cars such as Renault's Praxitèle at Saint Quentin-en-Yvelines near Paris, Fiat's Elettra Park in Turin; and Honda's City Pal in Motegi, Japan.

Car Sharing – Separating Ownership from Use

Leasing cars on a short-term basis, otherwise known as "car sharing," is another strategy aimed at reducing the impact of cars in cities. Car sharing gives urban residents access to cars without requiring them to own one. The concept works because members of car-sharing organizations do not depend on cars for everyday use. The typical member of an auto cooperative is a young, single, city dweller who needs personal transportation only occasionally. Car-sharing projects can generally be divided into three types: single-port systems (where users return the vehicle to the place where it came from), dual-port systems (to commute between two stations), and multi-port systems (where the use can leave it at any other port). Most existing car-sharing cooperatives are single-port systems. Multi-port systems remain technically challenging to implement because of the difficulty associated with keeping the vehicle offer in balance over the various ports given differing levels of demand across time and location.

Auto cooperatives have been multiplying rapidly in Switzerland, Germany, Austria, and the Netherlands. While early auto cooperatives appealed primarily to environmentalists and community activists wishing to declare their independence from the automobile, the current clientele is motivated more by personal convenience and cost savings than by ideological convictions. Car-sharing efforts in North America are still in their infancy, though fledgling car-sharing initiatives have sprung up in several cities in the United States, including Portland, the San Francisco Bay Area, Seattle, and Cambridge and in Quebec City, Montreal, Toronto, and Victoria in Canada.

Though car sharing is an interesting and innovative experiment, it is not currently expected to make a large reduction in the demand for personal automobiles in the industrialized countries. A study commissioned by the Swiss energy office estimates a market potential for car sharing of not more than 1.5% of the driving population.

<u>Fuel Taxes: Pricing Automobile Use Appropriately</u>

Appropriate pricing of the automobile as a tool toward achieving sustainability is a long-cherished goal of many economists. They argue that sustainability concerns arise because auto users capture all of the benefits of their trips, but pay only a fraction of the costs. In particular, drivers don't pay for the pollution, noise, and CO_2 they produce, the congestion delays they impose on other travelers, or the risk of accidents associated with their driving. Economists theorize that if drivers were asked to pay these costs through appropriate ownership and use charges, they would be more discerning in their travel choices. Lower and more sustainable levels of auto use in the aggregate would follow as a consequence.

Economists promote fuel taxes as a good (though not perfect) proxy for a "use" charge on gasoline use and consequently, for various pollutant emissions. The theory is that higher gas taxes influence consumer behavior in a myriad of complex ways. In the short term, consumers react by curtailing automobile use. The empirical evidence suggests that short-term effect is relatively minor – a 10% increase in fuel price translates to a 2% to 3% reduction in total automobile travel.

However, such automobile use differences understate the total impact of fuel taxes on sustainability. As the cost of gasoline consumption increases, consumers buy lighter, more fuel-efficient cars, thus reducing their gasoline consumption per kilometer traveled (though doing so causes somewhat of a rebound effect, as the impact of a higher fuel charge is reduced on a per-kilometer travel basis), and organize their lives (including where they live), in order to drive less. In turn, smaller and lighter cars make less of an impact if they are involved in accidents (though conversely their ability to protect passenger from other heavy vehicles is also reduced), and organizing housing decisions to drive less produces more compact suburbs and cities.

Indeed, empirical analyses of the effects of price on gasoline consumption in the OECD countries indicate that price increases have a very significant effect on gasoline consumption (and thus CO_2 emissions). Though the range of estimates varies significantly across studies and across different countries, the evidence suggests that a 10% increase in gasoline price has the effect of reducing total gasoline consumption by 6% to 8% with much of the reduction a consequence of consumers choosing to use relatively fuel-efficient automobiles. Gasoline prices in Western Europe are between two to three times the price of gasoline in the United States, and prices in Japan, Australia, and Canada are about 65% higher than in the United States. Further, analysis of 1995 data suggests that average fuel use per kilometer in the United States is about 50% to 80% higher than in Western Europe, and about 10% to 20% higher than in Japan, Australia, and Canada. Fuel use per capita in the United States is more than twice the per capita use in Western Europe and Japan, and about 70% higher than in Canada and Australia.

In other words, there is some evidence that where they have been implemented, high fuel taxes play a role in reducing gasoline consumption. However, increasing fuel taxes is a political challenge; such taxes are wildly unpopular with voters and have a regressive effect on poor and elderly drivers who live on fixed incomes. In most European countries, fuel taxes are already very high, and increasing them even more would be highly unpopular. This is also true in the United States, where fuel taxes are relatively low.

<u>Congestion Pricing</u>

Congestion pricing, or peak-period pricing, is a specific pricing scheme that charges auto users a premium for using road capacity when it is scarce, i.e., at peak periods. Singapore's "area licensing scheme," which has been in continuous operation since 1975, is often cited as an example of the potential effectiveness of congestion pricing as a tool to relieve congestion. Under this scheme, fees imposed on cars entering the city center during rush hours have resulted in roughly a 40% decrease in peak-hour traffic. Variable tolls implemented on several French auto routes on approaches to Paris and in three cities in Norway (Bergen, Trondheim, and Oslo) have likewise had a significant impact, by spreading travel demand to "shoulder" periods. However, efforts to introduce congestion pricing more widely have, thus far, met with only limited success. Until recently technology was a hurdle: The technologies needed to implement efficient tolling on high-speed, high-capacity roadways have become available only in the last decade. Furthermore, for a number of reasons, citizens and their politicians in most places have resisted the use of pricing to restrict peak-period driving. In the United States, attempts by the federal government to promote congestion pricing in the 1970s and again in the 1990s were rebuffed. Under a Congressionally authorized a "Congestion Pricing Pilot Program" enacted in 1991 and reauthorized in 1996, a number of communities carried out "pre-implementation" studies, only to conclude that there is not enough political support to proceed with implementation. Congestion-pricing initiatives in Sweden and the

Netherlands have likewise met with opposition. And attempt to implement a congestion-pricing scheme in London has also met with significant opposition.

Nonetheless, there are some indications that the future of congestion pricing is likely to be brighter than its past. First, technology is no longer a hurdle; the development and widespread experience with advanced electronic fare collection mechanisms renders the actual implementation of a congestion-pricing program relatively straightforward. Second, there are some recent experiences where congestion-pricing schemes have been successfully introduced without significant opposition. Politically, the best prospects for wider adoption of this strategy appear to be in connection with the introduction of new roadway facilities where the tolled facility offers a high level of service alternative to older, un-priced, competing facilities. In Southern California in the 1990s, implementations of this sort included the opening of a privately funded and operated SR91 Express Lanes system, allowing drivers of single-occupant vehicles to pay a fee to access a time-saving lane previously restricted to high –occupancy vehicles. Drivers still have the option of staying on the old, slower, but un-priced facility.

Enhancing the capacity and efficiency of the existing road and public transport infrastructure

Enhancements in the capacity or efficiency of infrastructure can address the economic sustainability of public transport and the operational sustainability of the road system.

Expanding the Physical Capacity of the Highway System

One way to relieve road congestion is to expand the capacity of the highway system by building new motorways, widening and improving existing arterial highways, and elimination bottlenecks caused by inadequate roadway design. Building new roads is the traditional approach to relieving congestion (or keeping congestion within tolerable limits).

However, it is clear that building new roads is not a foolproof "solution" for congestion. Building new roads is not always possible; it is almost always expensive; and the new capacity is regularly swamped by traffic growth induced in part by the anticipated easier travel conditions and changes in land use. Although it is difficult to measure "induced demand" – trips that would not have been taken but for the construction of the new roads – analysts agree that new roads induce some new traffic. Estimates of induced traffic's share of total traffic range widely, with estimates ranging from 10% to 100% of all traffic.

Building new roads is not as easy as it used to be. There is little reason to believe that opposition to new highways will lessen in the future. Although· new roads will be built, concerns about environmental and social sustainability will slow the pace of expansion of new infrastructure and widen the gap between travel demand and highway capacity to meet that demand. Hence, by itself, expanding road infrastructure does not seem to offer a valid strategy to achieve an operationally sustainable transport system.

Moreover, even discounting for new roads that will not be built because of community opposition, the demand for highways often exceeds the financial resources available. Thus new and innovative sources of financing are often key to successful infrastructure projects. The most prominent and popular of these new financing methods are new twists on an old idea – toll roads. Many new roads and highways in Europe, and some in the United States, are being built as toll facilities and are financed by the revenue from their tolls.

<u>Innovation to Increase the Operational and Economic Efficiency of Public Transport</u>

There are a number of initiatives afoot that promise to increase the operational and economic efficiency of public transport systems. Some are based on technological developments, such as the use of smart cards as a fare medium; the development of real-time passenger information systems that immediately inform passengers of delays in the system's extensive use of Global Positioning Satellites (GPS) – based automatic bus location systems; and dynamic scheduling and routing of paratransit to meet excess demand or make up delays. Still other innovative operational initiatives include door-to-door public transportation service.

Other initiatives involve more vigorous and imaginative management by public authorities and instructions. These include an increase in track sharing, i.e., joint use of mainline rail lines by intercity, regional, and municipal public transport system-wide; regional integration of public transport schedules and fares; the development of regional transportation associations; and the widespread outsourcing of public transport service operations to save money and improve service.

Among the most important global trends in public transport management are efforts to improve the economic viability and efficiency of public transport by turning the operation of public transportation systems over to the private sector. Known in its various forms as deregulation, privatization, outsourcing, contracting, franchising or competitive tendering, the aim is always the same: to improve the service quality and performance of public transport by injecting competition and entrepreneurial approaches into service delivery.

Deregulation and outsourcing are particularly popular with governments trying to improve public transport service. No longer associated with any particular political ideology, deregulation and outsourcing do not necessarily imply complete privatization. In many cases, the government retains policy-making functions (decisions about routes, fares, schedules and service standards) and contracts with private providers for service operation. Service contracts are awarded competitively to the lowest responsible bidder. In some cases, employees of a public transport agency may themselves compete for contract awards.

Operational Highway Improvements Using Intelligent Transportation Systems Technology

New and innovative management techniques and technology are helping improve the operations and efficiency of existing highways. Operational improvements are particularly effective in reducing unpredictable disturbances in the traffic flow caused by collisions, vehicle breakdowns, special events, and road repair. These improvements are made possible by recent advances in a set of communication and information technologies known collectively as Intelligent Transportation Systems (ITS) technology. The most significant technologies that have emerged after a decade of research and experimentation include:

Sensing and communication technologies. Three kinds of technologies are proving to be particularly important. The first are technologies that provide a means to detect disturbances in traffic flow quickly, enabling highway operators to clear accident scenes and restore normal operation conditions more rapidly. The second are vehicle-based emergency response systems using Global Positioning Systems (GPS) satellite technology, which can pinpoint a car's location even if the driver is injured or cannot accurately describe his or her whereabouts. A third application also uses GPS technology along with automatic vehicle identification techniques to monitor and control the movement and regularity of vehicles.

Advanced traveler information systems. Traveler information systems enable shippers, fleet operators, and individual travelers to make more informed transportation decisions based on real-time information about traffic conditions. For instance, real-time traveler information systems alert motorists to incidents and special events and advise drivers about alternative routings. Real-time parking information systems guide motorists to parking facilities that still have vacant spaces. Sophisticated roadway weather monitoring systems alert motorists to unfavorable road conditions and road hazards ahead.

Advanced payment mechanisms. Electronic toll collection (ETC) reduces bottlenecks at toll plazas, and makes possible the use of congestion-sensitive highway tolls that can increase charges during rush hours. Smart cards enable more efficient fare collection from transport riders.

Advanced traffic management technologies. Computer-controlled, traffic-responsive signal systems are used to smooth out traffic flow. Improved equipment and techniques shorten the time required for routine road repairs and reduce disruptions caused by work zones.

It remains to be seen whether intelligent transportation systems will have an impact on traffic congestion. So far, there is only limited anecdotal evidence of a positive impact of ITS on congestion as measured by improved traffic flow and reduced trip time.

ITS programs in the United States, Western Europe, and Japan have benefited from sustained government support. As these technologies demonstrate success and prove their worth, the need for continued government developmental support is being questioned. In the United States, there is an emphasis on transferring primary responsibility for implementing ITS systems to state and local governments. In Europe and Japan, national governments are expected to continue playing a central role in deploying ITS infrastructure.

Improving the Available Transport Options

Planners suggest two strategies to facilitate sustainable mobility in this category. First, reduce auto dependency by increasing non-auto transport option. Second, provide mobility and accessibility options for those who do not have access to autos.

Provision of Public Transport

In the last three decades, the cities of the developed world have significantly enhanced their public transport. In the EU, the bus and coach fleet has steadily grown and is now 50% larger than in 1970. The total number of buses in the United States has grown by over 80%, although most this growth is in the number school buses. In Japan, the bus fleet grew by about a quarter during this period. There has also been an expansion in urban rail in the last quarter century, with new systems constructed in a number of US and European cities. In some cities (for example, Atlanta, Miami, Naples, Bilbao, and Kyoto), it has taken the form of light rail. Most of the established metro systems outside the United States have been extended over the last two decades, and higher degrees of automation are being introduced. However, though these enhancements in the provision of public transport have been accompanied in most cases by increases in absolute levels of patronage, public transport's share of total trips and total kilometers traveled has actually declined almost universally across the developed world in this period.

Improving Non-motorized Transport

Among the OECD countries, Denmark and the Netherlands are the leaders in promoting NMT. The second Dutch Traffic and Transport Structure Scheme (SW2), covering the period 1990-2001, identifies the bicycle as the ideal mode for trips of up to 5-10 kilometers (in fact 40% of all automobile trips in the Netherlands are less than 5 kilometers in length). At the same time, the SW2 recognized a number of issues associated with bicycle use, including the need to provide direct, safe, and attractive bicycle routes between residences and trip destinations; the need to provide bicycle parking facilities; and the problems of safety and bicycle theft. Within the framework of the SW2, the government developed a national bike Master Plan (BMP), covering the period 1990-1997, to promote and improve bicycle use. Roughly 575 million guilders (US $230 million) was spent by central, provincial or local government on bicycle projects. Despite this substantial public investment, BMP research concluded that bicycle policy alone was not sufficient to increase bicycle use and restrain growth in car use.

Denmark has some of the most aggressive pro-NMT policies in the world. Copenhagen has roughly 300 kilometers of separated bicycle trails, roughly half the total length of the city's road network. Bicycles sharing the roadway with cars are given priority over turning vehicles at intersections, and a public education program inculcates a "culture of respect" for drivers, pedestrians, and bicyclists. Such initiatives have resulted (EU 2001) in Copenhagen's having one of the lowest rates of transportation-related fatalities per person in the world (1.3 deaths or serious injuries per year per thousand residents). Copenhagen also runs a City Bike program, which in 1997 provided roughly 2,500 free bikes at key locations around the city. The bikes are paid for by advertising, and are maintained by the Municipality, using prison inmates. There are plans to increase the number of bikes in the program. Copenhagen has also taken measures to make the use of cars undesirable; for example, it has reduced the availability of parking and converted streets to pedestrian zones. The city has also pursued a public transport-oriented urban development plan (the so-called finger plan, based on radial rail corridors). At the national level, automobile ownership in Denmark is discouraged through very high vehicle registration fees (105% to 180% of the vehicle purchase price), although the gasoline tax is in the middle range of European rates. Roughly one-third of the city's home-to-work trips are made by bicycle.

Providing Transport Options for Those Without Autos

There are many programs and policies to provide mobility for those without access to autos. Effective solutions frequently focus on particular groups, such as the poor, those with disabilities, or the elderly. Policymakers typically focus on three kinds of strategies:

Ensuring that mainline public transport services are sensitive to the needs of those without access to autos. Often agencies provide minimum levels of off-peak public transport service to ensure that service is available for so-called captive riders, even when such service would not be justified on strict economic grounds. Also, there has been a concerted effort across the developed world to ensure that public transport service is accessible to those with special needs. For instance, in the United States, the 1990 American with Disabilities Act laid out a minimum set of physical design standards for public transport vehicles and stations that public transport services must meet to serve those with special needs.

Paratransit services. In several regions, there are trips that conventional public transport is unable to serve. In many cases, local authorities provide dedicated demand-responsive paratransit services to help people with special needs complete those trips. Such programs are often targeted at the disabled and the elderly. (Examples include "The Ride", offered by the Massachusetts Bay Transportation Authority in the Boston metropolitan region.) These services have both advantages and disadvantages relative to conventional public transport. Being demand-responsive and door-to-door, they often offer a high level of service, and as a result there is disagreement about what constitutes fair and efficient pricing for the service. On the other hand, these services usually require significant advance planning. Also, many citizens with disabilities argue that as a matter of dignity they deserve to be integrated with mainstream society as much as possible, and being able to use public transport is an important element of this effort. Many citizens with disabilities and their advocates decry targeted paratransit services – even those offering higher levels of service than conventional public transport – as humiliating.

Direct user-side subsidies to help those without autos either get them (in the case of the poor) or buy alternative transport services (such as taxi service) directly.

Land-Use and Urban Design Strategies

In the last three decades, some urban regions in the developed world have successfully employed land-use policy to facilitate a pattern of development in which public transport can play a significant mobility role, and limit sprawl.

Before discussing alternative policy regimes, it is important to note that land-use policy works best in rapidly growing residential and commercial developments where existing structures can be easily reused over time. In established regions, the durability of exiting housing stock limits the effectiveness of land-use measures. In the United States and the United Kingdom, for example, annual construction of new housing represents only around 1% of the existing stock.

Public Transport-oriented Development

This policy encourages residential, employment, and recreation buildings to cluster around rail public transport stations. The goal is to build compact, pedestrian-friendly communities where many trips can be made on foot or by bicycle, and longer commutes by train.

This approach has been followed widely – and successfully – in Europe, Canada, and Japan. Examples include Stockholm's satellite towns along commuter rail lines radiating from the city; the French villes nouvelles on the outskirts of Paris; Toronto's high-rise developments along the Yonge Street subway; Japan's public transport communities at the termini of private rail lines; and Germany's transit-oriented suburban communities such as Munich's Perlach and Frankfurt's Neustadt. In the United States, on the other hand, only a few examples of public transport villages exist. In a great many suburban locations, community opposition to high-density development makes public transport-oriented development difficult to implement.

Spatial Location Policies: The Dutch ABC Policy

The Netherlands, one of the smallest developed nations in Europe, has a comprehensive approach to land-use planning: the ABC policy. Dutch planning focuses on curbing traffic growth and urban sprawl, and also, on developing compact cities and protecting open areas. ABC explicitly seeks to reduce automobility through programs such as the one encapsulated in it's slogan for business location: "the right business in the right place."

The ABC policy classifies businesses into three categories based on the importance of their need for public access and road transport. Business development sites are similarly classified in terms of their public transport and road accessibility. The policy attempts to encourage businesses with a high number of employees and visitors to locate on sites with good public transport accessibility, such as near centrally located public transport or rail stations ("A" sites) or near major public transport nodes in less central location ("B" sites). "C" sites, with good road access, are primarily intended for businesses that depend on road transport for their operations. Associated with each type of site are restrictions on the number of parking spaces that can be provided there: "A" sites are limited to 10 to 20 parking spaces per 100 employees. These rules are sufficiently restrictive that businesses have a strong incentive to locate in accordance with the intentions of the policy.

Overall, the Dutch accept the ABC policy, through objections to the highly restrictive parking limits, along with economic pressure at the local or provincial levels, have led to a relaxation of the parking rules in some areas of the country.

Integrated Approaches

The most successful examples of cities controlling automobility and improving the sustainability and their transport system use combinations of the policy options above. Isolated policy responses are unlikely to have a significant impact.

Copenhagen, for example, combined public transport-oriented land-use planning, high automobile ownership charges, priority treatment of bicycles, and numerous improvements to center-city social life. Portland applied spatial growth controls, development of light-rail public transport system and, again, many enhancements to improve the quality of center-city life.

As another example, Zurich upgraded its tramways into a modern, high-quality, and reliable public transport system operating on separate rights-of-way obtained by removing traffic lanes from general use. A computer-based signaling system ensures that trams do not have to stop at traffic intersections. Seasonal "Rainbow" passes reduce riders' cost per trip to very low levels. Intensive marketing and information campaigns promote the use of the tram system, and special maps show people how to get to particular destinations such as restaurants and cultural attractions via the public transport system. These public transport system improvements were accompanied by complementary land-use and urban improvement policies. Large shopping centers were developed around major stations. Urban public transport villages were developed around the public transport lines, including Tiergarten, a public transport village built in an abandoned quarry in central Zurich. As a result of these and related measure, the automobile mode share for the journey to work fell by 9% between 1980 and 1990.

Sustainable Practices in Industry[9]

This reading includes selected case studies, which illustrate that sustainable practices in industry go far beyond technical solutions, and address cultural, social, economic and community needs as part of sustainable management practices. They are grouped into seven categories: Innovation & Technology, Eco-Efficiency, Managing and Understanding Change, Dialogue and Partnership, Providing and Informing Customer Choice, Corporate Social Responsibility, and Creating

---

[9] This reading draws upon a series of case studies collected by the World Business Council for Sustainable Development. More information is included in (Schmidheiny et al. 1997), and available at http://www.wbcsd.org.
This material is included in this report with permission.

Sustainable Livelihoods. One illustrative case study from each category is included in this reading.

Innovation & Technology:
"Zero" Targets Driving Innovation at DuPont

Founded in 1802, DuPont is a science company, delivering science-based solutions from operations in 70 countries with 83,000 employees. In its 200-year history, the company has undergone several transformations, evolving from an explosives company to a chemical company and now to a science company.

The DuPont mission is to achieve "sustainable growth" which is defined as creating shareholder and societal value while reducing footprint throughout the value chain. Paul Tebo, vice president for safety, health and environment has been a driving force behind implementing sustainable growth within DuPont. Tebo has been spreading the vision to DuPont businesses worldwide, setting challenging targets based on elimination of all injuries, illnesses, incidents, wastes, and emissions throughout the value chain. In short, 'The Goal is Zero'. The critical aspect of the goal is that businesses must still grow while driving towards zero.

The Goal is Zero impacts each of DuPont's core strategies- from improving productivity, to increasing knowledge intensity and finally to delivering new products through integrated science. The mission of sustainable growth is creating alignment between business strategies and societal expectations and the Goal of Zero is driving new innovations within the company.

Innovations include progress on reducing waste and emissions at DuPont sites. A global team developed new technology for the manufacture of Terathane brand PTMEG, a key raw material for Lycra. The innovation increased yields, resulting in additional revenues of $4 million while eliminating 4.4 million pounds of waste per year.

Another team developed and implanted methods to reduce approximately three million pounds of annual releases of HFC-23 through process optimization. The innovation saved $20 million in capital investment and reduced greenhouse gas emission on a CO_2 equivalent basis by 40 billion pounds. In Asturias Spain, the Sontora business determined that second quality material could be used productively rather than be waste. With the assistance of DuPont and some other local organizations, a group of unemployed women formed Novatex S.A. to take the second quality Sontora and produce one-time use products for medical and laboratory applications. Novatex is now a stable business with 13 direct and stable jobs for women who previously had difficulties being hired into the local economy while material that was formerly a waste is now a valuable product.

The real benefit to growth has been in those Goal of Zero innovations that have gone beyond DuPont sites to include customer impacts. A Packaging and Industrial Polymers team in Europe created a peelable lid system for packaging application that eliminates solvent emission from lacquer coatings. This innovative product reduces this packaging material and improves tastes and odor impartation. As a result of this effort, DuPont has gained a 10 percent share of the lidding market and reduced more than 1,000 tons of methyl acetate solvents per year in Europe. A DuPont of Canada team instituted a new business model with Ford Canada. Instead of selling gallons of paint, DuPont sold painted cars. Over the four-year term of the program, Ford's emissions were reduced by 50%. A Crop Protection team established a vision to help the poorest people in the world to continue to grow cotton with less risk to their safety and health. According to official figures in Benin alone, 37 people died from misapplication of older type of cotton insecticides. In looking to enter this new market, the West African DuPont team developed a safe product that reduced application rates by a factor of 10, designed appropriate packaging, and then trained officials, distributors, and farmers on the safe use of the product.

While the sustainable growth transformation and the attainment of zero goals throughout the whole value chain will take time, already the zero challenge is driving new innovations through the business globally. Many DuPont teams are beginning to recognize the opportunities that meeting zero targets presents, turning sustainability challenges into business opportunities. Similar examples are emerging from every DuPont business. The challenging targets are forcing DuPont businesses to rethink products and approaches and come up with new innovative solutions to drive the sustainable growth transformation. For more information on DuPont's progress toward sustainable growth see DuPont (2004).

Eco-efficiency:
Water Conservation and Reuse Program at the Ramos Arizpe Auto Complex (RAAC) of General Motors de México

General Motors de Mexico (GMM) Ramos Arizpe Auto Complex (RAAC) is located in an arid region in the State of Coahuila, in northeast Mexico. The complex opened in 1980 in Ramos Arizpe (pop. 40,000) and by 2000, RAAC was manufacturing 590,000 engines, 171,000 transmissions, and assembling 222,000 passenger vehicles.

The only source of water in the area where the RAAC is located is a small, semi-confined aquifer that has a limited water storage capacity and a relatively high salt content (0.2%). This does not allow direct well-water use for industrial domestic purposes. Since 1986, several things have occurred that changed RAAC's approach to water management. These include:

Well-water levels have decreased

Fees that must be paid to the National Water Commission (Comisión de Aguas Nacional – CAN) for water rights have substantially increased

CAN imposed limits on well water withdrawal

Limits on the concentrations of several parameters in the waste streams were issues, and

The demand for high quality water has increased, due to the expansion of the RAAC

Targets

The company's challenge included:

Securing water for production processes without depleting the aquifer (which is also the local drinking water source)

Desalinating the well water supply

Establishing a recycling and reuse process for the industrial and sanitary wastewater – all within the framework of an intensive water conservation program

Actions

RAAC has undertaken several programs to reduce water consumption, to suppress pollution due to industrial and sanitary wastewater discharge, and to reuse treated effluents.

Specifically these programs are:

The development of a continuous intensive water conservation program that included leak detection and repair, and a review of the different water-treatment and water-reusing processes to detect water-saving opportunities

The implantation of an innovative system to recover most of the by-product brine from the reverse osmosis systems. This would increase the portion of reusable water from the systems thereby reducing the amount of well-water withdrawn and extending aquifer life

The construction of solar evaporation ponds to convert the final brine stream to solid salts for potential reuse. This avoids discharge of high levels of salt into a stream/creek system that is ultimately used for crop irrigation purposes

The implementation of a complete physical-chemical and biological wastewater treatment facility to treat all industrial and sanitary wastewater

The implementation of an innovative system to recover about 70% of the secondary effluent that results from the biological treatment of a pre-treated industrial wastewater

The reuse of treated sanitary wastewater to irrigate RAAC gardens and sports-fields and to create a man-made lagoon. This area is the center of a recreational area for workers and families.

Results

Water conservation activities during the early days (1986-1997) show a steady reduction in the amount of water taken from the wells. Predictably, the subsequent years of conservation activities show fewer reductions as the "big-hits" came in the earlier period.

The Brine Recovery Systems (BRS) has been operating almost continuously for about 3 _ years. It has produced water with low salt content that has been used mainly for the engine and transmission plants' water supply. It has also allowed RAAC to substantially reduce the well-water withdrawal through an enhanced, more efficient use of well water by increasing the useable portion of water withdrawn from the well from 67% to 94%.

As a result of employing solar evaporation ponds, RAAC has been able to avoid the discharge of a salt-loaded stream to a creek whose water is used for crop irrigation purposes. The industrial and sanitary wastewater treatment has also reduced the pollution load of these streams, and prepared them for recovery through the BRS.

The integrative solutions have been a success for both the region and the production complex, which employs nearly 6,000 local residents and now produces annually some 590,000 engines and 222,000 passenger vehicles, mainly the Chevrolet Cavalier, Pontiac Sunfire and Chevy 3-, 4- and 5-door pickups.

Through these efforts, GM Mexico has reduced annual well water withdrawal from 1,470,000 m3/year in 1986 to 700,000 m3/year in 2000, reducing the average amount of well water needed to produce a vehicle from 32 m3 to 2.2 m3. At the same time these reductions took place, production jumped – the company now manufactures about 7 times more cars, 50% more engines and brought on-line a new transmission plant that manufactures 171,000 unites in the year 2000.

The programs described in this document are an integral component of RAAC's Environmental Management System. Papers describing RAAC RO by-product brine and industrial secondary effluent treatment and recovery were presented at the 1998 and 1999 International Water conferences held in Pittsburgh, Pennsylvania, and the Weftec-2000 Conference held in Anaheim, California. Numerous awards have been presented to RAAC by both the Mexican government and ISO 14000 certification officials in recognition of the Complex's environmental accomplishments.

Managing and Understanding Change: A Learning Experience at Interface

Interface, Inc. is the world's largest manufacturer of carpet tiles and upholstery fabrics for commercial interiors. But Interface's core vision is not about carpet or fabrics per se; it is about becoming a leading example of a sustainable and restorative enterprise by 2020 across five dimensions: people, place (the planet), product, process, and profits. As a company with 27 factories, sales offices in 110 countries, annual sales of US $1.3 billion, over 7,000 employees and supply chain heavily dependent on petrochemicals. that is a substantial challenge.

Founder and Chairman Ray Anderson presented this challenge to the organization in 1994. As a result, Interface has undergone considerable transformation in its effort to reorient the entire organization. Some positive results were achieved in the beginning. Through its waste elimination drive, the company has saved $165 million over five years, paying for all of it sustainability work and delivering 27 percent of the group's operating income over the period. Over and above that, since 1994 Interface, Inc. has reduced its "carbon intensity" – its total supply chain virgin petrochemical material and energy use in raw lbs per dollar of revenue – by some 31 percent. However, Interface recognized that sustainability means far more than that.

The company developed a shift in strategic orientation based on a 'seven-step' sustainability framework, using the systems thinking of The Natural Step. These steps include elimination of waste (not just typical waste. but the whole concept of waste); elimination of harmful emissions; use of only renewable energy; adopting closed loop processes; using resource-efficient transportation; energizing people (all stakeholders) around the vision; and redesign of commerce so that a service is sold that allows the company to retain ownership of its products and to maximize resource productivity.

Throughout the business, all employees were trained in the principles of systems thinking. They were required to examine the impact of their work and how they could work more sustainably in their business area. The feedback on this training has been very positive and a great deal of progress has been made as a result. However there were three areas where Interface could have improved the process.

The first is always establishing a positive environment for inspired employees, fresh from their training courses, to return to. The company found that employees became passionate as their understanding of sustainability grew, and they need an outlet for action. Although there were many areas of good supportive management across the business, there were also too many areas where local managers were not prepared well enough to facilitate motivated employees wanting to make a difference. Issues of management and leadership explain why some of the expected progress did not happen in certain areas.

Second, people engage in different ways with sustainability issues, and learning programs need to provide the space to explore these differences. Programs need to be flexible enough to go into detail on a 'hot' issue such as climate change, while the next question may well be about equity of resource use. To keep people motivated, programs need to maintain this flexibility.

Third, follow up was not quick enough; it takes much more than two days for people to really understand sustainability. Sustainability issues need to be revisited again and again, as employees being to understand how it impacts their daily lives. It is a big commitment to revisit these issues on an ongoing basis, but the company recognized that it was vital for employees to continually buy in.

Interface has also learned the importance of making sustainability a 'whole company' approach. Those who 'got it' quickest were inevitably those working in either the manufacturing or the research areas of the business, people used to talking about the environment, systems, and material substitution. A high number of the company's early wins came on the manufacturing side: green energy purchase, waste elimination, and recycling. However, it took longer to really achieve the buy-in of the sales and marketing teams, with the result that 'whole company' issues such as strategic product development, planning and communicating sustainability externally took longer to be integrated.

For Interface this has been a comprehensive sustainability learning approach for the company; a great deal has been learned along the way and the process has benefited from mistakes and successes. The company is now very aware that sustainability needs to become "business as usual" for everyone across the business, and these experiences have been a solid contribution to successful change. The company is now well on the course to achieve its 2020 vision. For more detailed information on Interface's sustainability program, please consult Interface (2004).

Dialogue And Partnership:
Water Use Planning at BC Hydro

BC Hydro, one of the largest electric utilities in Canada, is a Crown corporation owned by the province of British Columbia. BC Hydro generates about 50,000 gigawatt-hours of electricity annually, 90% of which is clean, renewable hydroelectricity. BC Hydro serves nearly 1.6 million BC customers, while the power subsidiary, Powerex, markets energy products and services in western Canada and the western United States.

In response to ever-increasing competing demands for the province's abundant water resources, Water Use Planning (WUP) was developed. This program is a joint initiative of BC Hydro, and the British Columbia (provincial) and Canadian (federal) governments, in consultation with First Nations and the public. The overall goal is to find a sustainable balance between competing water uses that are socially, environmentally and economically acceptable.

The dams and reservoirs used to store and regulate water at BC Hydro's 30 hydroelectric facilities affect fish and wildlife habitat, cultural resources, recreation activities, and water levels. In addition to power production, these facilities create other benefits such as domestic drinking water supply, flood management, and economic development.

A Water Use Plan (WUP) is a detailed set of operational instructions for a facility, focusing on the timing and amount of water releases through various dam release structures. Water Use Plan decisions at each hydroelectric facility are being made within an open, inclusive multi-stakeholder committee designed to consider economic, social and environmental values.

Committee members include federal and provincial government agencies, BC Hydro, First Nations, local citizens and other public interests. In developing a Water Use Plan, participants apply their values and interests to the evaluation of different operating alternatives, seek multi-party consensus on a preferred alternative, and make recommendations regarding water management at BC Hydro facilities.

Because effective, sustainable water use is fundamental to BC Hydro's success, the company is highly committed to the WUP process. Water Use Plans will prove clear and understandable boundaries within which to operate and to plan for future electricity supply. Senior government and the BC Comptroller of Water Rights will do final WUP reviews and consultations.

Targets

The main objectives of the WUP:

Support BC Hydro's corporate mission to provide integrated energy solutions to customers in an environmentally and socially responsible manner

Help meet BC Hydro's objective of building and maintaining public support by engaging external constituents, including other government agencies, the public, First Nations, our regulators and our shareholders, in a dialogue about options, tradeoffs and priorities in operating our hydroelectric facilities

Reduce or eliminate regulatory uncertainty by setting clear boundaries that take into account public values. This makes good business sense because it provides operational stability

Apply societal values into the use of water for multiple interests and objectives

Enable more informed decision-making that transparently incorporates public values into tradeoffs between water use interest, opportunities and impacts

Actions

A core group of representatives from BC Hydro and the provincial and federal governments developed the WUP framework over a two to three year period. This Inter-Agency Management committee continues to guide and support to the WUP program.

The WUP program has been underway since 1999, with 2 plans completed, 13 underway and the rest scheduled to begin within the next year. Each plan follows the 13 step WUP Guidelines. The key features are designed to:

Set interest-based objectives and related performance measures

Identify relevant information or research studies

Create alternative operating strategies

Make values-based tradeoffs

Monitor WUP outcomes

Review and document both the procedure and the decisions

Through the WUP process, stakeholders from different organizations and the public have been collaborating on new studies, interpreting data results and making science-based decisions that reflect a broad range of societal values.

Many WUPs will have an ongoing monitoring component, taking an adaptive management approach that blends scientific investigation and value-based decision-making to seek the sustainable balance between economic, social and environmental interests.

Over five years the BC Hydro WUP program's costs are approximately $26M CDN, with key provincial and federal partners supporting the program through staffing and other resources. As WUPs find a sustainable balance between economic, social and environmental interests, there is a likelihood that ongoing scientific monitoring and reduced operating flexibility will decrease the financial value of power generation. The provincial government, BC Hydro's sole shareholder, will recognize the costs of implementing Water Use Plans through reductions in water rental payments from BC Hydro to the province. The cost estimate for implementing Water Use Plans at the outset of the program was $50M CDN. Although rough, this estimate provided government with the comfort to approve and embark on the WUP process, with some sense of the magnitude of costs and a general idea of the benefits for non-power values.

Problems and Difficulties

The Challenge: The WUO program is faced with strict budgets and timelines within which it must carry out extensive public consultation for the 30 hydroelectric facilities. This has presented numerous challenges at times; such as limited money for new research, and competition for limited human and monetary resources between water use plans, by all participants. The WUP process has approached these challenges in several ways:

- o Solution – Develop expertise and learning in the decision analysis, environment and procedural areas that creates benefits for subsequent WUPs. Earlier WUPs had larger budgets and timeframes; later WUPs benefit from their predecessors' experiences, in planning, research and staff and participant experience

- o Solution – Adopt an "adaptive management" approach where key uncertainties in the science or the baseline data are addressed either through a comprehensive monitoring plan, or through an "active" water use plan that implements flow trials in an experimental design to study the impacts of different operating scenarios

o   Solution – Have different "levels of intensity" for different facilities. This allows larger budgets and more time where more science, debate and conflict are anticipated and a "slimmer" process for simpler facilities with fewer impacts and possible operating alternatives

o   Solution – Collaborate with other agency participants in the development of the WUP program schedule. This helps participants to schedule their own resources and priorities

The Challenge: WUPs are multi-disciplinary, multi-stakeholder decision processes that do not fit neatly into the current business-planning paradigm. The diversity of perspectives, areas of expertise and values within the company offer both challenges and opportunities with a multi-disciplinary program.

o   Solution – WUP project teams balance "WUP-dedicated" staff (environmental modeling, resource valuation and project management) with expert staff from facility operations, corporate planning and Aboriginal relations. This team structure allows the WUP program to develop and grow, while capitalizing on the existing planning and operations expertise in the public, values-based process to enable multi-directional learning.

The Challenge: Earning the trust and respect of both agencies and the public, and gaining a multi-year commitment of time, energy and resources, required BC Hydro to demonstrate its commitment to the WUP process.

o   Solution – Earlier projects were assigned larger budgets, ensuring time and flexibility to address participant needs. Both the process design and BC Hydro earned trust, through shared learning and exploring diverse values. BC Hydro now confidently leads the process, following the WUP Guidelines while benefiting from tools and learning form earlier WUPs

o   Solution – Alongside WUPs, BC Hydro initiated the annual $1.5M CDN Bridge Coastal Restoration Program, supporting eligible fish and wildlife projects within these watersheds to address issues related to dam construction and as a predecessor to the WUP program. The Bridge Coastal program augments the existing Columbia and Peace Fish and Wildlife Compensation Programs (annual allotments $3.8M CDN and $1.5M CDN respectively).

The Challenge: Securing and supporting the widespread participation of First Nations in WUPs was recognized from the outset as both a key to success, and a potential challenge. First Nations governments and communities in British Columbia have varying levels of expertise and interest in participating in a WUP process. Furthermore, First Nations are participating in a great number of decision-making processes that further tax their resources and create competing priorities.

Solution – The First Nations (FN) WUP Committee has a representative on the Inter-agency WUP Management Committee, as well as on the other program development activities. First Nations participants in WUP can access the FN WUP Committee for assistance in WUP participation, including scientific and legal advice concerning aboriginal rights to resources.

- o Solution – The multi-attribute tradeoff process offers an effective and equitable tool for incorporating traditional ecological knowledge (TEK) into "western" decision-making. Cross-cultural learning is a valuable objective of all WUPs.

Results

As an effective route to interest-based negotiation and structured decision-making, the WUP process is widely supported and accepted because it offers superior results to all participants, including BC Hydro, over traditional positional or litigative dispute resolution

Explicit learning about values and interests of all participants. BC has been a leader in multi-stakeholder land-use planning, and participants at WUP tables understand and embrace the benefits of interest-based, multi-party decision making. Participants tend to move from black and white thinking toward a greater understanding about the interplay between economics, social and environmental interests. The complexities of BC Hydro's integrated operations are becoming better understood, while BC Hydro learns about the intricacies of fisheries and habitat management and social values. The end result is more informed decisions and shared responsibility for outcomes

Inter-agency and inter-personal relationships are built through the WUP process, between government agencies, First Nations, the public and BC Hydro. This will be an ongoing legacy of WUPs, making collaboration and negotiation a preferable alternative to cooperation from agencies, rather than the opposition and conservative environmental management that are often sought when the science and effects are unknown or uncertain

The scientific database, methodologies and transferable knowledge are built through investment in sound scientific investigation into the operational effects on fish, environment, recreation and First Nations' values. A WUP may refine, remove or add requirements and operational effects. Scientific modeling and methods will have ongoing value and benefit to BC Hydro and other hydroelectric facilities.

The completed Stave Water Use Plan achieved incredible results. Recommendations for change were made through the WUP process and once adopted, these changes met with improvements across all objectives, including economic, social and environmental objectives. They type of win-win situation clearly demonstrates that collaboration can lead to sustainable decision-making for all players.

New hydro-operations will reflect social and environmental costs. Program costs will be assessed in several ways: traditional calculation of economic losses due to restricted power generation will be balanced against reduced litigative costs, and gains in economic welfare and economic efficiency by internalizing environmental and social impacts

BC Hydro's Water Use Planning process is ongoing. Many of the lessons learned at early WUP tables are being applied to the next round of tables.

> Inclusive representation in the WUP process that allows for a broad range of public interests. This ensures credibility, exposure to the full range of ideas and options, wide spread support for the process and an effective and efficient WUP program

> Explicit and transparent sharing of information, interests and values is essential to completing a successful WUP, and to ensuring participants contribute in good faith through to the end

> Creative exploration of alternatives has led to participants seeking ways to achieve mutual benefits, rather positions that exclude or complete with other interests

> Focus time and the process on the learning and creative steps. Large group processes can run the risk of using up "precious time" on procedural issues – effective facilitation through all of the WUP steps enables participants to spend time actively working and learning in the steps they care about

For up-to-date information on the BC Hydro Water Use Planning program, including links to individual Water Use Plan projects and a "quick facts" section, please consult BC Hydro (2004). As of August 1, 2001, 2 WUP tables are completed, 13 are in progress and the final 7 are scheduled to start in the next year.

Providing and Informing Customer Choice: The Climate Neutral Network

A handful of companies are going beyond "energy efficiency" to seek ways to leave no carbon footprints behind. Their Climate Neutral Network stretches the horizons of orthodoxy for those companies that are trying to reduce greenhouse gas (GHG) emissions.

Over the past three years, the US-based Climate Neutral Network has built and alliance of companies that are learning how to build market share and customer brand loyalty by offering their customer's products and services that achieve a net zero impact on the Earth's climate.

Participating companies can become certified as Climate Cool on achievement of complete reduction and offset of all carbon emissions. A company that chooses to become a Climate Cool enterprise agrees to reduce and offset all of the climate impacts for the full spectrum of its internal operations from the point at which raw materials are received to the point at which finished product is delivered. Products or services can also be certified individually as Climate Cool. Product certification requires a reduction and offset of the greenhouse gases generated at each stage of the life cycle on a cradle-to-grave basis: the sourcing of materials; manufacturing or production; distribution, use, and end-of-life disposal.

Why do companies participate? Shanklee Corporation, the first company to receive Climate Certification, see participation in the Network as an opportunity to leverage the company's 40-year history of environmental focus and performance. Participation is a means of branding the organization and not just the individual products. "We felt strongly about moving beyond past performances and striving just for reductions" says Ken Perkins, Environment, Health & Safety Director. "We were attracted by the bold objectives of being climate neutral. It is not just old wine in a new bottle; the Network is innovative, groundbreaking and distinguishable".

Organizations in the network are now actively collaborating to co-design new Climate Cool products and partnerships, and a creative and rapidly expanding company-to-company market is developing. The Saunders Hotel group, also certified Climate Cool, was seeking lighting accommodations. Saunders needed to develop a means of reducing the energy consumption of hotel lighting without sacrificing ambiance. Philips Lighting worked with Saunders to designs alternative bulbs. They came up with more energy-efficient alternatives that were also smaller in size and weight.

"This provided an opportunity to plug-in new designs," says Paul Walitsky of Philips Lighting, "the climate Neutral network is providing new business opportunities to put products in practice".   Philips Lighting is also working with US apparel firm Norm Thompson to provide energy-efficient lighting to the Ecumenical Ministries of Oregon. As a result of this and other efforts, Norm Thompson recently received their Climate Cool certification.

The Network more recently began certifying events that achieve a zero emission footprint. This includes a comprehensive calculation of the estimated emissions from an event suing the Climate Neutral metrics that closely mirror the WBCSD and WRI Greenhouse Gas Protocol Corporate Accounting and Reporting Standard.

The first certification of this kind was the 2002 Winter Olympic Games in Salt Lake City. This includes measurement of greenhouse gases from such sources as athlete, official and spectator travel to Salt Lake City, transportation around town, and the events and venues themselves, including even the burning of the Olympic torch.   The offsets required were achieved through donations of greenhouse gas reduction from DuPont and Petro Source, and the effort was supplemented by the planting of 18 million trees. The reductions that DuPont is donating are from process related reductions of nitrous oxide, representing reductions beyond their internal commitment.   In total, the emission reductions exceed the calculated emissions footprint of the Games by three times.

"By becoming Climate Cool, companies are issuing a leadership challenge," says Sue Hall, executive director of the Climate Neutral Network.   "These exciting precedents are attracting broad interest from many companies who are able to leverage Climate Cool activities for their own objectives."

For more information on the Climate Neutral Network see CNN (2004).

Corporate Social Responsibilty:
Eskom HIV/AIDS Program

Eskom is South Africa's wholly state-owned electricity utility. It has 24 power stations and is one of the lowest cost producers of electricity in the world. The company supplies electricity to over three million customers via 306 thousand km of transmission and distribution lines, with a nominal capacity of 40,585 megawatts. Eskom supplies approximately 95% of the country's electricity and is one of the lowest cost producers of electricity in the world.

Since the start of the AIDS epidemic 83% of all AIDS deaths so far have taken place in Africa. The sub-Saharan part of Africa holds the majority of the worlds infected individuals. In South Africa, the estimated prevalence is currently 4.7 million individuals. As an organization conducting its business mainly in South Africa, the containment and management of HIV/AIDS is a strategic priority for Eskom.

Targets

Eskom's employees, suppliers and customers were all important components of Eskom's HIV/AIDS program. The main objective of the program was to minimize the impact of HIV/AIDS on Eskom, and thus a set of programs were formulated and implemented throughout the organization, and resources were dedicated to these programs.

Information Management – Establish infrastructure to help Eskom maintain a strategic focus on the developments related to the prevalence of HIV/AIDS in the business

Self-awareness – Increase the level of self-awareness of HIV status among Eskom employees

High Risk High-risk areas and situations for contracting HIV Infection were identified in the organization, and are being addressed by various business units.

Communication – Develop and implement a communication strategy for the overall response strategy to support the strategic management of HIV/AIDS

Education and Training – To empower all employees with skills, knowledge and information to deal with HIV/AIDS effectively

Care and Support This program caters for psychological support of HIV positive employees, the free treatment of sexually transmitted infections and monitoring of TB treatment

Policies and Practices – The program ensures that Eskom policies and practices do not discriminate against HIV positive

Actions

The actions taken in order to meet the targets were:

A dedicated budget of 125 South Africa Rand per employee was spent on HIV/AIDS projects and activities in 2000 (excluding the salaries of employees working full-time on the HIV/AIDS program)

Partnerships were established with national and international institutions and organizations working in this field. This includes a search on a HIV/AIDS vaccine where Eskom works with the Department of Health, Medical Research Council and Institute of Virology

Benchmarking Eskom's activities against South African businesses and International businesses Eskom also conducted a voluntary, anonymous and unlinked surveillance study in 1999. Following these results, Eskom commissioned the Harvard Institute for International Development (HIID) to analyze the short, medium and long-term economic impacts of HIV/AIDS to the organization, in order to enable it to scale up its interventions. A set of Response Strategies was formulated to meet the challenge

During 2000 a cycle tour was arranged to raise funds to fight the HIV/AIDS epidemic. An education and awareness program was launched. This included a wellness week, induction programs for new employees and peer education workshops

Participation of Eskom's business units in World AIDS day with various programs

Results

All Eskom employees have received HIV/AIDS awareness messages, so that the challenge is now "beyond awareness"

Eskom is a member of the Global Business Council (GBC) against HIV/AIDS

Eskom contributed 30 million South African Rand over 5 years towards the vaccine development

Eskom chairs the Southern African Power pool forum on HIV/AIDS. The main purpose of this forum is to share experiences and assist in capacity building

Eskom shared its experience and assisted more than 20 companies in the country with information to help them start their own programs

Achievements

Eskom's HIV/AIDS program was presented to the XII World Aids conference, and was commended as a good model for a workplace program

Eskom's HIV/AIDS program has received two international awards, for Business Excellence by the Global Business Council, and for the involvement of people with HIV/AIDS in the programs, including employing them as members of the staff, from The United Nations AIDS program

The HIV/AIDS Program was nominated for a South African award and has three best series documents that are distributed world-wide

Lessons Learned

Businesses are not immune to the devastating impact of HIV/AIDS

New infections were projected to cost Eskom 4-6 time the annual salary per individual infected

Annual costs of existing HIV infections during the years 2006-2010 will average 7% of the payroll

Creating Sustainable Livelihoods:
Farmer-sized Seed Packs in a Sustainable Community-Oriented Development Program

Kenya is importing food. Its farmers, despite their skills, cannot produce enough to feed themselves and the urban populations. Soil has been depleted by years of continuous cropping, and is chronically deficient of nutrients. Thus crop growth is poor, and farms lack organic matter. In some parts of Kenya, despite good rainfall, corn now grows only knee-high, leaving the ground exposed to rain and erosion. Yields are as low as 50 kg grain/ha, far below the potential yield of four metric tons/ha.

State and international aid programs have provided fertilizers, ploughs, and tractors on credit. But when the programs end, farmers are unable to pay back the loans and continue to maintain the equipment. The size of provisions has also been a problem, with the smallest bags of fertilizer often weighing 50 kilos. "This is an investment of a few weeks of income, and you can only carry it around if you have a bicycle," says Paul Seward, and plant nutrition expert. Seward has teamed up with

Dismas Okello, the local community development expert who formed the Sustainable Community-Oriented Development Program (SCODP). Over the last five years, they have worked closely with farmers in Siaya district to devise a mini-pack project to provide quality seeds and fertilizers at affordable prices.

"For 5 shillings (US 6 cents) you can buy a chewing-gum sized sachet of 250 seeds of the vegetable Sukuma-wiki. For another 10 shillings (US 12 cents) you get a pack of fertilizer for 150 planting holes. The results are dramatic. A good farmer can earn anything from between 2,000 to 4,000 shillings (US \$25-50) from using these packs." Having sold its first kilo bag of fertilizer five years ago, SCODP is now selling 300 tons of fertilizer a year to small-scale farmers in affordable quantities. The market potential is for several hundreds of thousands of tons.

SCODP has grown since 1997 to become a well recognized and well-supported, successful grassroots projects. Specialist assistance has been provided to SCODP to address the many pests affecting corn, sorghum, beans, cowpeas, and pigeon peas, including training in crop protection related issues. Field trials were started in 2000, and are continuing in order to develop solutions for specific pest problems. Crop protection products in small packs and small application devices are now also part of the whole technology package that farms can access through SCODP. Due to its success, USAID and the Rockefeller Foundation are assisting SCODP to extend its approach to other regions in Kenya.

Engineering and Construction Research for Sustainable Development[10]

A report prepared by the Civil Engineering Research Foundation (CERF), entitled "Creating the 21st Century through Innovation; Engineering and Construction for Sustainable Development," provides 38 research prospectuses in five construction industry categories: Management and Business Practices, Design Technology and Practices, Construction and Equipment, Materials and Systems, and Public and Government Policy (CERF 1996). A selection of those prospectuses most related to sustainability and sustainable development from each of the five categories is included in this reading. The overview and challenge for each category is also included.

While somewhat dated, this reading is included in this volume because of its global innovation and comprehensive research agenda for achieving sustainability in

---

[10] This reading is abstracted from the 1996 Report of the Civil Engineering Research Foundation (CERF), entitled "Creating the 21st Century through Innovation; Engineering and Construction for Sustainable Development." (CERF 1996) This material is included in this report with permission.

the construction industry, much of which still needs to be accomplished in advancing sustainable engineering practice.

Management and Business Practices

An effective management process is a key ingredient in providing the design and construction services needed for sustainable development in the 21st century.

"Management," defined as the act, art, or manner of planning, organizing, controlling, or directing, is by its very nature a diverse topic. It can be applied to almost any human activity. Management skills have always been particularly critical when it comes to the successful completion of construction projects with their complex and diverse elements. Add to this the demands of sustainability, and the requirement for strong, effective management becomes even greater.

The Challenges for Management

The burden of satisfying both the need to efficiently deliver safe, reliable, and functional infrastructure, while preserving and enhancing environmental quality, is the responsibility of construction industry management. To meet this challenge, management will need clear, uniformly understood and accepted objectives that define the parameters of sustainability and new tools and methods that allow the monitoring of the process. Without this basic information, management will be unable to make sustainable practices a reality.

Numerous barriers also impede the adoption by the construction industry of innovative technologies that support sustainable development. For example, the construction process all too often is marked by adversarial relationships and distrust that make consensus building difficult—if not impossible—to achieve. Existing procedures for project selection, characterized by inflexible and inappropriate rules and regulations and codes and standards, also inhibit management's ability to use innovative technologies and systems to further sustainable principles. Finally, given the dynamic nature of the industry, management often lacks knowledge of state-of-the-art alternatives to apply innovative solutions to a particular project.

Prospectus Summaries

Managing Sustainability in Project Selection

Owners of constructed facilities usually consider direct costs and regulatory requirements in selecting constructed facilities. Often ignored are other considerations such as renewability, recyclability and reuse, energy consumption, operating cost, waste pollution reduction, and risk and liability. These should be incorporated into an owner's decision process. This prospectus proposes to field test and conduct demonstrations to provide information based on the operation and maintenance cost of these facilities to help owners become familiar with sustainable construction opportunities that ultimately increase the likelihood they will be adopted.

<u>Establishing Affordable Sustainable Construction Objectives</u>

This prospectus addresses the cost impact of achieving sustainability. The work should develop a framework for allocating costs and benefits, develop methods for observing these values, and report on the implications for design, construction, materials, operations, and human resources. Anticipated outcomes of the research will include a model for conducting cost-benefit analyses that involve sustainability. These models should include examples and case studies, an electronic tutorial for distribution on the Internet, and applications across various projects in the public and private sectors.

Design Technology and Practices

Design is the process of identifying and solving problems having to do with the form, use and management of facilities and communities, infrastructures, and other elements of our built environment. Design encompasses a range of activities that precede construction or other changes in the built environment. The outcomes of design inevitably depend on the context in which they perceive and represent problems and solutions.

The process and outcomes of design are central to achieving sustainable development. Private and public sector entities act on the designer's recommendations to change and use the built environment. Design technology and practices—the procedures, methods and tools designers use—shape the ways designers perceive problems and solutions as well as influence designers' productivity, accuracy and efficiency.

The Challenges for Designers

Design Technology and Practices industry experts see the need for defining sustainability, for streamlining the design process, for developing knowledge-base systems, and for bringing understanding and awareness of sustainable design and construction opportunities. These themes encompass the challenges facing designers seeking to achieve design goals and to lead the engineering and construction toward achieving sustainable development in the 21st century.

Designers are critical players to the successful introduction of innovative and sustainable facilities. The challenges facing the industry broadly are often encountered first by construction designers. Meeting these challenges requires dealing with a series of barriers, including a public and the design community that lack a strong background on sustainable development design opportunities; a limited design knowledge base that forces designers to rely heavily on personal experiences; a design process that is not conducive to innovation owing to the lack of involvement of key stakeholders; and the lack of a clear understanding of sustainable design principles, which makes pursuing innovative efficient design difficult.

Prospectus Summaries

Sharing the Sustainable Design Knowledge Base

Designers currently rely heavily on their own personal experiences. This situation exists in part because of a lack of readily accessible information on global design practices and experiences. This prospectus proposes to address this deficiency by establishing an international clearinghouse, through the collaboration efforts of the design community and government, for the exchange of information on design principles and practices.

Forming Consensus Performance Criteria for Sustainable Facilities

There is a growing appreciation of the need for society to account for sustainability in their actions. There is no consensus that defines sustainability for facilities that feature political, social, economic, and technical considerations. This prospectus recommends that the industry convene a working group of key stakeholders to establish consensus on sustainability in facilities and to develop the methods for defining, benchmarking, measuring, monitoring, and disseminating results. Participants should commit to share relevant information, which would be managed by a lead organization. Results would be disseminated to industry.

Establishing Education and Training Programs to Advance Sustainable Design

This prospectus recommends the development of basic design education that incorporates the multidimensional nature of design for sustainability. Design education will address broader scope and engage the full range of stakeholders to become more effective. This work would develop specific collaborative education programs and initiatives through a multiple strategic alliance of educators, practitioners, professional associations, and funding sources. The expected outcome of the prospectus would include a curriculum for education in sustainability; the engagement of multiple constituencies from precollege through the professional level, to public education; and a specific pedagogical and technological infrastructure. And international conference on education for sustainability, involving practitioners, academics, and accrediting bodies should be held to address its implementation.

Construction And Equipment

Improved construction and equipment technologies and practices are needed to meet the goals of engineering and construction for sustainable development in the 21st century. Improved construction and equipment will safely clean-up hazardous sites, protect the safety of workers, increase their productivity, improve the quality of the constructed facility, and reduce the time required to bring facilities into service.

In seeking sustainable development for the 21st century, construction will be the area of action in which practical innovations will be rewarded and impractical concepts eliminated, and the goals of sustainable development will succeed or fail. Construction and equipment technologies and practices must, by necessity, be very well defined and thought out prior to execution. Specific technologies must be developed and deployed.

The Challenges for Construction

Construction as the front line of the engineering and construction industry faces the challenges of staying abreast of innovative technological developments and creating tools and methods that enhance the industries competitive position, dealing with a process that is often hostile to innovation, and seeking to advance innovation through practical means such as demonstration projects.

Prospectus Summaries

Achieving Sustainable Development through Sharing Knowledge between Developed and Developing Countries

Developing countries are experiencing rapid growth, with the associated need for their infrastructure to support that growth. At the same time, developed countries are replacing old and obsolete infrastructure to meet existing and future requirements. Technologies should be developed to meet the social, economic, and environmental needs of all these citizens. Technology transfer will be a large factor in this process. Recognizing that we can each learn from each other, this prospectus proposes to develop construction and equipment strategies that help transfer infrastructure innovation. In the near term, attention would be given to identifying existing technologies that best meet the needs of developing countries and conducting demonstration projects in cooperation with host countries. The results will be disseminated to decision makers of developed and developing countries. In the longer term, the world would investigate research and demonstration activities that would facilitate the transfer of innovative technologies.

Developing a Decision Support System for Eco-Construction

Traditional life-cycle assessments of construction projects lack an explicit consideration of the environmental impact of alternative approaches. This prospectus involved an international effort to develop a life-cycle framework that explicitly incorporates environmental impacts into the assessment of construction activities. This effort would be coordinated among national construction industry associations and other interested bodies. Current methodologies and practices would be shared with the ultimate aim of building consensus eco-construction methodologies.

Developing an Integrated Sustainable Project Life-Cycle Management Approach

The compartmentalization and fragmentation of the project-delivery process have led to lack of accountability among participants in the various projects phases (i.e., planning, design and engineering, manufacturing and procurement, construction, operations, maintenance, repair, demolition, recycling, and waste treatment), and corresponding suboptimization within the overall project. This prospectus proposes to reduce the barriers to integration by a variety of measures including promotion of strategic alliances, rationalizing the regulatory and legislative structures, using government projects to set the example, developing and overall government policy framework to encourage integration (e.g., as is done with the design-build approach), shifting to performance-based specifications and standards, maximizing the use of information technologies, developing appropriate incentive structures, and improving decision tools such as the use of life-cycle benefit analysis.

Materials and Systems

Materials and Systems are a key strategy in developing contemporary building and civil infrastructure solutions that will support sustainability objectives. The industry must not only develop high-performance materials and systems, but it must also promote global action plans to expedite their technology transfer and application. This strategy requires the adoption of life-cycle cost minimization principles, including environmental impacts, since new materials and systems that facilitate design durability, ease of construction, demolitions and reuse directly benefit the life-cycle characteristics of the built environment worldwide.

The Challenges for Materials

Materials experts recognize that the availability of appropriate materials and the development of high-performance materials and systems are critical to enhanced sustainability. Achieving the benefits from advanced materials and systems requires overcoming barriers. For example, there may be conflicts between using high performance materials and systems and the ability to recycle or to reuse materials and systems. Likewise, the appropriate application of high-performance materials and systems to address sustainability differs both by geographic region and by type of construction, i.e., new construction, maintenance, repair, replacement and upgrades.

Prospectus Summaries

Establishing a Materials Life-Cycle Information System for Constructed Facilities

Sustainable development practices will depend on the availability of reliable life-cycle assessment capabilities that currently are not adequate. This prospectus centers on performing a worldwide public-private effort to develop models and methods that identify and gather data needed for life-cycle analyses, assessment of materials and systems, and renewal assessments. Industries advanced in life-cycle methods, such as the chemical, electronics, and communications industries, will be recruited as collaborators. The result of this effort would be comprehensive and systematic life-cycle information systems for the engineering and construction industry.

Establishing Training Programs for Life-Cycle Cost Analysis

The engineering of sustainable facilities requires evaluation of alternatives for materials, systems, and construction-operation approaches. This prospectus proposes to develop the tools and databases to support these analyses of alternatives and create training and education programs targeted at professionals, and a global communication network to reach the professional community and public policy and opinion leaders as well as professions. The objectives are expected to be realized through a combination of symposia, application of Internet communication capabilities, the development of model curricula, dissemination of existing databases and creation of new ones, collecting case histories of applications, and publishing a journal targeted to the issue.

Public and Government Policy

Public and Government Policy construction industry experts concentrated on three interrelated government policy topics: the role of engineering and construction officials in sustainable development, public/private mechanisms to meet performance and sustainability goals, and promotion of innovation. The group believes that the ability of the construction industry to realize long-lasting and large-scale change is limited without concomitant change in public and government policy. The three policy tops are seen as underpinning the industry's desire to effect change and the Symposium's goals and vision.

Policy team members recognize that their effort must include not only the construction industry members, who by their actions can influence policy, but those who will use the recommendations to shape public and government policy. Member of this target audience exert enormous influence on project funding as well as the enactment of laws and regulations. Within this context, the construction industry profession has a critical role to play in enabling needed public and government policy reform.

The Challenges of Policy

Historically, the construction industry has remained disengaged from public and government policy making. There are many reason for this, but chief among them is that the engineer, public work official, academician, and others who make up the construction industry profession, generally tend to favor involvement in technical level activities with their peers rather than involving themselves in the unscientific business of policy development and implementation. Some observers have referred to the "disempowered condition" of the civil engineering profession because of their silence and lack of leadership on this issue. The important role this industry will play in determining the success of sustainable development requires that it get involved.

Prospectus Summaries

Using Field Testing to Demonstrate Sustainable-Development Practices

To achieve sustainability, development projects in the 21st century will move from a parochial disciplinary focus to a multidisciplinary approach that seeks mutual understanding, coordination, and consensus. In order to show how this multidisciplinary approach can be organized and be effective, this prospectus proposes to target one or more complex projects (such as the rehabilitation of urban brownfields) where engineering and community interests are paramount. Multidisciplinary partnerships would be examined, and models for success would be developed and tested at other sites. At the macroscale, the prospectus proposes that a multidisciplinary coordinated approach be pursued involving key construction stakeholders to advance the sustainability "profession," to promote innovation and collaboration on sustainable development, and to encourage new national and international approaches to sustainable development.

Taking a Leadership Role in Defining Sustainable Solutions

The public opinion leaders and policy makers who make up the constituency of the design and construction industry are the ultimate decision-makers in implementing change. This prospectus recognizes the need to provide more information on the issues and consequences of actions to that constituency, and proposes an organized program of public education to accomplish it. In particular, the proposed effort would be implemented through a broad context of issues (ranging from prospects for the future to specific design approaches), across a variety of mass and targeted media (TV, radio, Internet, community meetings, and showcase displays in public arenas, and so on). The information would be designed with a multidisciplinary focus, and include case studies of success stories to highlight benefits. The program would be guided through the development of coalitions and partnerships coordinated internationally, but integrated to focus on the delivery of targeted programs at the local level.

<u>Expanding the Construction Industry Knowledge Base on Sustainable Development</u>

Local, national, and international engineering societies and associations have the potential to reach the over ten million engineers in practice worldwide. This prospectus proposes to align and focus the engineering and construction community through development of a process to network key organizations (professional societies and associations) for the purpose of sharing information on sustainable development in practice, education and research. The goal of this prospectus is to use existing professional societies and associations to address the development and implementation of sustainable development principles in engineering and construction. A task committee should be established that would identify and contact all groups that have or should have an interest in sustainability as it applies to construction. The task committee would work toward the establishment of permanent joint committees to carry on the collaborative work.

The Greening of Chemical Engineering[11]

This reading has been selected as an illustration of how a major engineering discipline and organization is addressing the challenges of sustainable engineering practice.

Making EHS (Environmental, Health, and Safety) an Integral Part of Process Design

This book presents an approach – termed MERITT (Maximizing EHS Returns by Integrating Tools and Talents) – for enhancing process development through better integration of environmental, health, and safety evaluations. It draws upon critical components of inherent safety, pollution prevention, green chemistry, and related paradigms through selective adoption and adaptation of their existing tools, skills and knowledge resources.

---

[11] This material is from a book published by the American Institute of Chemical Engineers (AIChE) in 2001, entitled "Making EHS (Environmental, Health, and Safety) an Integral Part of Process Design" (AIChE 2001). It also includes material from initiatives of the Center for Waste Reduction Technologies (CWRT), an affiliate of the AIChE, on Sustainable Development and Environmental Stewardship. (CWRT 2004). Reprinted with permission. Copyright © 2004 AICHE. All rights reserved.

MERITT is based on five "C" Fundamental Principles

1. *Commitment* – There must be commitment from the process development leaders, the project managers, the business managers, and the corporate leadership. This is ultimately a top down requirement that must support and fit within the corporate culture for positive reinforcement of the principles, rather than conflict.

   Clearly articulate business value through success (and failure) stories as well as examples. Ensure that the individuals' metrics as well as corporate reward and recognition systems also reinforce desired behaviors.

2. *Concurrency* – Thinking about environmental, health, and safety concerns concurrently while identifying issues, setting priorities, and defining and actualizing opportunities is the cornerstone of the approach. Concurrency is all about *fusion* and *infusion*—fusion of environmental, health and safety perspectives, and infusion into the development process. Several of the other principles directly bolster concurrency.

   Draw from existing in-house joint Environmental Health (EH), Environmental Safety (ES), Health Safety (HS), and Environmental Health and Safety (EHS) projects and practices, such as in facility auditing and life-cycle assessment work, as examples of how these disciplines can successfully work together. Emphasize the criticality of getting the right input in a timely fashion.

3. *Communication* – Both concurrency and collaboration require communication. Throughout the development process there must be open and ongoing communication to get the right mix of expertise and knowledge involved at each step of the development process; exchange ideas and resolve issues; and disseminate information to all active participants, business leaders, and customers.

   Focus communication efforts on the important issues. Make them valuable and short. Do not inundate people with day-to-day developments and "newsy" narratives.

4. *Collaboration* – Coordination of activities is clearly necessary but not itself sufficient in achieving concurrency. There must be collaboration. Individuals must work together in identifying, addressing, and resolving EHS requirement. It is important to note that collaboration does not require formal meetings. Often smaller working groups are much more successful.

Collaboration (and buy-in) can be successfully effected through the simple act of joint development and/or use of resource components, such as tools or metrics.

5. *Continuity (Concept to Completion)* – Considerations and evaluations must start at the earliest possible stages of the project and the effort must be carried through successive development steps. In essence, a *continuum* for organized thought and information transfer must be created that will lead to informed decision making at each step so that each successive effort can build upon what has already been done. By the same token, up-front work must anticipate the future needs in the development process for the work to proceed efficiently. This necessitates collaboration.

Avoid "upstream" barriers. Getting "downstream" development people involved earlier in the process can provide preemptive feedback that can help avert information gaps or precluded options later. Similar experience from other projects can also help.

Many leading companies have already embraced most of these principles within corporate environmental and/or safety programs; however, most have yet to achieve complete success in the areas of collaboration and continuity. There are even fewer companies, if any, that fully address the integrated EHS perspective. Progress in these areas will require both dealing with a company's formal institutional structure (e.g., organizational format, management style, and information systems) and contending with the company's culture and customs (e.g., unwritten rules and legacies of entitlements). An effective ongoing training program is generally regarded as essential to support changes of management systems, procedures, or behaviors.

Center for Waste Reduction Technologies (CWRT)

Established in 1991, the mission of the Center for Waste Reduction Technologies (CWRT 2004) is to "benefit industrial sponsors and society by leveraging the resources of industry, government, and others to identify, develop, and share technology and management tools that measurably enhance the economic value of industrial sponsor organizations while benefiting the environment or addressing issues of sustainability."

The CWRT recently published a Strategic Report on tools for sustainable development and environmental stewardship. The Report states that "more and more, businesses have entered into a new phase in which sustainable business performance is regarded as: a global competitive advantage; a catalyst for innovation; and a way to capture new market and financing opportunities. New technologies are needed to allow us to maintain economic growth while decreasing material and energy intensities. Investing in sustainable technologies is expected to lessen the extent of future requirements for remediation, and monitoring and control. CWRT recognizes this leading-edge issue by developing technologies and management tools that can help in their drive towards sustainability."

The Center has brought together key players from the chemical, pharmaceutical, petroleum, and downstream manufacturing industries, as well as their suppliers and technology contractors, to do collaborative work. Among the collaborative projects included in the Report are the following:

1. *Total Cost Assessment Methodology* – The Total Cost Assessment (TCA) methodology provides a disciplined and standardized approach to identifying all life-cycle costs and benefits associated with decisions related to environmental health and safety (EH&S) issues for industrial products and processes. Prior to the development of this tool, an industrially accepted approach to conducting a TCA was not publicly available.

   The methodology, which has been beta-tested by CWRT sponsors, captures the costs associated with manufacturing operations for users and manufacturers of chemicals from raw material extraction to ultimate ecological fate (the entire life cycle of a process or product) and helps decision-makers to assess the total costs incurred, cost savings accrued, and the cost avoided for materials, products processes or services that explicitly include the costs of EH&S issues.

   The methodology is not intended to replace a corporations' existing financial accounting system. Instead, it is intended to be a decision support tool that aids the user in making informed decisions regarding EH&S opportunities and impacts.

2. *Combining the Principles of Inherent Safety, Pollution Prevention, and Green Chemistry in the Design of Sustainable Products and Processes* – Historically, safety and pollution prevention disciplines have been practiced independently, but there is now growing awareness that a methodology that could combine the two disciplines and include concepts from "green" chemistry would offer a powerful new paradigm for the design of new processes and products and the improvement of existing ones.

   Many manufacturing processes (both new and retrofit) are often designed in a sequential "stagegate" manner with little integration between technical disciplines or EH&S, reviews in the developmental processes. This project is directed towards developing a validated methodology that can be overlaid on exiting practices. The project methodology will help to create a more integrated, multi-disciplined, life-cycle approach to design, which is expected to result in significant design improvements throughout the development cycle and benefits in plant operation. Coupling the chemistry research phase to life cycle, EH&S, and 'manufacturability' issues early in the development phase is a key concept in this methodology.

3. *Baseline Metrics for Sustainability* – A study group of twelve major companies in the chemical, paper, pharmaceutical, and materials industries has been working since late 1999 to identify impact categories and associated metrics that would be useful tools for industry. The group's objective is to develop, pilot, and benchmark metrics that complement the work of the World Business Council for Sustainable Development and others, with a focus on metrics useful to CWRT sponsor companies. Metrics have been developed covering material and energy intensity, pollutants (greenhouse gases, photochemical ozone creating gases, acidification, and eutrophication), water, human health, and eco-toxicity. These metrics are all normalized against value added, net sales, and per unit of product. The study group is currently testing these metrics within their businesses, with the objective of refining the base metrics and identifying other complementary metrics which can help drive continual improvement in both business value and level of sustainability.

4. *Sustainability Metrics for Industry* – In late 1999 two projects were started to test and refine the energy and material metrics for small- and medium-sized enterprises that are manufacturers and suppliers of chemicals and their customers, and to test the metrics along key product supply chains (PVC products, ABS pipe, nylon rope, and carpet underlay). The data sources being mined in this work include the U.S. DOE's Industrial Assessment Center Database, Carnegie Mellon's Environmental Input/Output Lifecycle Assessment Database, and the European Plastics Industry Lifecycle Assessment Database. In addition, eco-efficiency metrics are being calculated for the 50 most important products in the U.S. chemical industry while evaluating the feasibility of extracting complementary indicators using SRI's PEP (Process Economics Program) data. Selected results from this study will be tested against current industrial practice to validate this approach.

Global Warming and Climate Change in Minnesota

This reading is included as an example of public information outreach by a State on a major sustainability issue (MDOC 2004): the Minnesota Department of Commerce, Office of Environmental Assistance, and Pollution Control Agency.

A Big Experiment

On global basis, we know that in recent years the surface of the earth is warming. Nine of the ten warmest years in the instrumental record (1861-present) have occurred since 1990. 1998 was the warmest year on record and 2001 was the second warmest. Most scientists agree that the earth's surface temperature warmed more during the last century than any other century during the last thousand years (WMO 2001). There is new and stronger evidence that most of the warming observed for the last 50 years is attributable to human activities. Recent warming appears to be linked with the burning of oil, coal and gas for energy in vehicles, businesses and homes and increased atmospheric levels of carbon dioxide.

Impacts of Global Warming

Global warming is changing our climate and is impacting our natural systems. While it is impossible to say that any one event is due to global warming, the following evidence is consistent with what would be expected in a warming climate.

> Sea level has risen worldwide approximately 15-20 cm (6-8 inches) in the last century.

> A general increase in heavy rainfall events contributes to increased flooding and erosion.

> Coral reefs are dying from the warming of tropical seas. Some experts predict many will be gone by 2002.

> Annual lake and river ice cover declined by about two weeks in the Northern Hemisphere during the past century.

> 50 million acres of Alaskan forest are under attack by spruce budworms that thrive in warmer weather.

> In Glacier National Park, the number of glaciers has fallen from 150 to 50 since 1850.

Has Minnesota's climate changed over the past 100 years?

The average temperature in Minnesota has risen almost one degree Fahrenheit over the past century.

Since 1900, precipitation has increased by about 20 percent in parts of Minnesota, especially southern Minnesota.

If temperature readings and precipitation continue to increase within the next century, Minnesota might soon feel and look more like Missouri.

What can we expect?

The following impacts may already be underway or could be expected from a changing climate.

Shifts in location of Minnesota forests and grasslands, changing the types of plants and animals that live in the state.

Loss of species unable to adapt quickly to new climates.

Damage may increase from floods and violent storms.

Some pests, diseases and exotic species may be able to extend their range into Minnesota.

More poor air quality (smoggy) days in the summer.

Reduced water quality – increased algal blooms and less oxygen in warmer waters.

Less habitat for trout, whitefish, and other coldwater species.

Shorter season of snow and ice cover, less winter recreational opportunities.

What Can You Do To Help?

Dealing with the impacts of global warming will require the involvement of everyone: governments, businesses and individuals.

Steps to reduce carbon dioxide emissions will reduce the main greenhouse gas contributing to climate change. In addition, our actions can also save money and protect our air, water and soil.

Six things you can do:

1. *Purchase green power* – Minnesota law requires the state's electric utilities to offer customer voluntary options to purchase power generated from renewable sources that emit far fewer greenhouse gases. Contact your electrical provider for details.

2. *Reduce your energy use for transportation:*

   o Buy a fuel efficient or alternative fuel vehicle

   o Drive less – bus, bike, walk or carpool

   o If you do drive, don't idle your vehicle

3. *Reduce energy use at home:*

   o Turn your thermostat down in the winter, set your air conditioner higher in the summer

   o Buy energy-efficient bulbs and appliances; look for the new ENERGY STAR label

   o Get a home energy audit

4. *Plant trees* – Trees capture and hold carbon dioxide, a major greenhouse gas, and provide shade, which can reduce the need for air conditioning.

5. *Reduce, reuse, and recycle* – Waster reduction and recycling saves energy and resources.

6. *Educate others* – Share the facts on global warming and encourage all to do their part.

St. Paul Environmental–Economic Partnership Project: Working for a Sustainable Community

This reading is included as an excellent example of the continuing commitment of a metropolitan community with a history of innovation in the realm of environmental preservation. A summary of the Project follows.

Municipalities need not choose between a healthy and environmentally sound community on the one hand, and a prosperous community on the other. We can have both.

The Environmental-Economic Partnership Project was initiated in 1993 to implement the City's Urban CO_2 Reduction Plan.

Project Goals

Saint Paul City council pledged in 1992 to reduce CO_2 emissions 20% below 1988 base year levels of 5,700,000 tons/year by 2005 = 2,393,000 tons/year (20% of projected 2005 emissions), with an intermediate goal of 7.5% by 1997 = 899,000 tons/year (=Kyoto)

"To encourage present activities and identify future activities that improves both the environmental and economic health of Saint Paul."

"To foster actions for sustainable development, including energy retrofits in municipal buildings, recycling and waste reduction, equipment/lighting conversions, water treatment efficiency changes, land use planning and neighborhood development, district heating/cooling and transportation improvement."

Note: *All CO_2 reduction calculations/numbers derived using ICLEI Greenhouse Gas Emissions Software by Torrie Smith Associates

Project Objectives

Public Health and Safety

Reduce local air pollution levels

Preserve public health

Relieve traffic congestion and prevention of further urban sprawl

Fiscal Responsibility

Reduce energy bills

Reduce waste and future clean up costs

Reduce maintenance & operating costs

Strengthened Economy

Increase employment

Create new marketing opportunities

Lower operating costs

Reduction in energy bills for government, business, and families through increased efficiency

Promoting the City

Prevent further urban sprawl

Create amenities that draw people

Local, national and international attention

Projects

Energy

District Cooling Service

Hot Water District Heating and Cogeneration

Wood-burning Biomass Plant

Sewer Pumping Station Improvements

Street Lighting and Signal Lamp Conversion

Cities for Climate Protection Campaign

Sustainable Buildings Task Force

Climate Wise Program

Energy Efficiency Policies

Conservation Improvement Project Building Retrofits

ENERGYSTAR

Pumping Electricity Real Time/Peak Demand Pricing

Transportation

City Employee Metropass Program

Bike Lanes and Paths

Mayor's Commuter Action Team

Traffic Calming

Downtown Computerized Traffic Signal System

Traffic Signal Re-timing Program

During Incidents Vehicles Exit to Reduce Travel Time (DIVERTT)

Advanced Parking Information System (A.P.I.S.)

Dale-Maryland-White Bear Avenue CO Reduction Project

Forestry & Parks

Greening of the Great River Park

Citywide Tree Planting Program

Wood Recycling Programs

Water Quality

Mississippi Riverfront Corridor Plan

Storm Water management Program

Sewer System Rehabilitation Program

Wetlands Restoration

Treatment Chemicals Reduction

Lime Sludge Dewatering

Control of Lead in Drinking Water

Neighborhoods

Polluted Lands (Brownfields) Cleanup Program

Empowerment Zone/Enterprise Community

Phalen Corridor Redevelopment Initiative

Phalen Wetland Restorations

Recycling & Waste Management

Yard Waster Compost Sites

Curbside and Multi-family Recycling Programs

Neighborhood Clean-Up Program

City Building Recycling Programs

'Free Market' Residential Waste Exchange

Waste-to-Energy Facility

Environmentally Friendly Purchasing Policies

CO_2 Reduction Plan Strategies

Strategy	CO_2 target reduction	Reduction achieved so far
Strategy #1 – Municipal Action Plan	10,800 tons	19,200 tons
Strategy #2 – Diversification of the Transportation Sector	731,000 tons	Unknown
Strategy #3 – Urban Reforestation	3,600 tons	2,100 tons
Strategy #4 – Energy Efficiency	1,354,000 tons	276,000 tons
Strategy #5 – Energy Supply	283,200 tons	84,000 tons
Strategy #6 – Recycling and Waste Prevention	10,800 tons	75,300 tons
Overall E-EPP Project CO_2 Reduction per year in 2001 = 460,000 tons per year = 20% of 2005 goal of 2,393,800 tons CO_2 reduction = 56% of 1997 intermediate goal of 899,700 tons (exceeds current Kyoto objectives).		
Total cost savings per year = $41,000,000		

Primary Project Partners

District Energy Saint Paul, Inc. (DESP)

Private, non-profit, community-based corporation founded in 1979.

DESP operates the largest hot water district heating system in North America.

75% of downtown Saint Paul's building space, including State Capitol Complex and 298 single-family residences.

District Cooling Saint Paul, Inc., began providing cooling service in 1993. Currently 45 buildings in Saint Paul.

Converting to 100% biomass-generated power from burning 280,000 tons/year of clean wood waste at cogeneration facility downtown Saint Paul; supply 25 MW to the Xcel Energy grid beginning 2002.

District Energy provides energy management and conservation assistance to downtown area building owners.

For more information on District Energy, please consult DESP (2004).

Xcel Energy (formerly Northern States Power/NSP)

City's Conservation Improvement Project (CIP) financed by NSP through no-interest loans of up to $1 million per year for five years began in 1993.

Savings produced by the improvements are roughly equal to the city's loan payments.

Improvements pay for themselves within 10 years or sooner.

Xcel benefits by avoiding need to build new power plants and by reducing peak demand at its existing plants.

Xcel signed power purchase agreements with District Energy Saint Paul to purchase 25 MW of biomass-generated power beginning in late 2002.

Xcel subsidiary NRG Inc. operates 1,000 ton per day Ramsey/Washington County refuse-derived-fuel facility, which burns Satin Paul solid waste at Xcel power plants for electricity.

Xcel has CIP projects with many other entities in Saint Paul.

For more information on Xcel Energy, please consult Xcel (2004).

Saint Paul Neighborhood Energy Consortium (NEC) and Neighborhood Recycling Corporation/Eureka Recycling

Coalition of community-based organization that provide environmentally responsible community information, services & programs.

Eureka Recycling operates curbside and multi-family recycling programs for City's 110,000 households.

City is working with NEC to increase energy audits and provide insulation and conservation advice for homes, apartments, and businesses.

For more information on the Saint Paul Neighborhood Energy Consortium, please consult NEC (2004).

For more information on Eureka Recycling, please consult Eureka (2004).

Chapter III
Sustainability Definitions, Policy Statements, and Principles

This chapter provides a selected sampling of sustainability definitions, policy statements, and principles, with explanatory notes, that will give the reader a broad sense of the visionary and ethical goals and strategies of sustainable engineering practice. It also contains a wealth of additional information that can be used to pursue the basic material in the Report in greater depth and detail.

Selected Definitions of Sustainability

The following is a sampling of definitions of sustainability that have been crafted by various organizations over time. They have been selected to provide the reader with a historical and visionary framework for understanding the philosophical dimensions of sustainability and sustainable development.

World Commission on Environment and Development (WCED) – 1987

"Sustainable development is a process of change in which the exploitation of resources, the direction of investments, the orientation of technical development, and institutional change are all in harmony and enhance both current and future potential to meet human needs and aspirations...

Sustainable development... meets the needs of the present without compromising the ability of future generations to meet their own needs."

International Federation of Consulting Engineers (FIDIC) – 1990

"Development that will meet the long term needs of future generations of all nations without causing modification to the Earth's ecosystems."

Institute of Engineers, Australia (IEA) – 1994

"Sustainability is the ability to maintain a high quality of life for all people, both now and in the future, while ensuring the maintenance of the ecological processes on which life depends and the continued availability of the natural resources needed.

Sustainability is *the ability to maintain a desired condition over time.* Sustainable development *is a tool for achieving sustainability,* not the desired goal."

President's Council on Sustainable Development (PCSD) – 1996 (Sustainable America)

"Our vision is of a life-sustaining Earth. We are committed to the achievement of a dignified, peaceful, and equitable existence. A sustainable United States will have a growing economy that provides equitable opportunities for satisfying livelihoods and a safe, healthy, high quality of life for current and future generations. Our nation will protect its environment, its natural resource base, and the functions and viability of natural systems on which all life depends."

Selected Policy Statements on Sustainability

The following policy statements on sustainability have been selected to provide perspectives from the business community, the consulting community, and two major engineering organizations.

Declaration of the Business Council on Sustainable Development (BCSD) – 1992[12]

Business will play a vital role in the future health of this planet. As business leaders, we are committed to sustainable development, to meeting the needs of the present without compromising the welfare of future generations.

This concept recognizes that economic growth and environmental protection are inextricably linked, and that the quality of present and future life rests on meeting basic human needs without destroying the environment on which all life depends.

New forms of cooperation between government, business, and society are required to achieve this goal.

Economic growth in all parts of the world is essential to improve the livelihoods of the poor, to sustain growing populations, and eventually to stabilize population levels. New technologies will be needed to permit growth while using energy and other resources more efficiently and producing less pollution.

Open and competitive markets, both within and between nations, foster innovation and efficiency and provide opportunities for all to improve their living

---

[12]   This declaration was prepared by the World Business Council for Sustainable Development. More information is available at: http://www.wbcsd.org.
This material is included in this report with permission.

conditions. But such markets must give the right signals; the prices of goods and services must increasingly recognize and reflect the environmental costs of their production, use, recycling, and disposal. This is fundamental, and is best achieved by a synthesis of economic instruments designed to correct distortions and encourage innovation and continuous improvement, regulatory standards to direct performance, and voluntary initiatives by the private sector.

The policy mixes adopted by individual nations will be tailored to local circumstances. But new regulations and economic instruments must be harmonized among trading partners, while recognizing that levels and conditions of development vary, resulting in different needs and abilities. Governments should phase in changes over a reasonable period of time to allow for realistic planning and investment cycles.

Capital markets will advance sustainable development only if they recognize, value, and encourage long-term investments and savings, and if they are based on appropriate information to guide those investments.

Trade policies and practices should be open, offering opportunities to all nations. Open trade leads to the most efficient use of resources and to the development of economies. International environmental concerns should be dealt with through international agreements, not by unilateral trade barriers.

The world is moving toward deregulation, private initiatives, and global markets. This requires corporations to assume more social, economic, and environmental responsibility in defining their roles. We must expand our concept of those who have a stake in our operation to include, not only employees and shareholders but also suppliers, customers, neighbors, citizens' groups, and others. Appropriate communication with these stakeholders will help us to refine continually our visions, strategies, and actions.

Progress toward sustainable development makes good business sense because it can create competitive advantages and new opportunities. But it requires far-reaching shifts in corporate attitudes and new ways of doing business. To move from vision to reality demands strong leadership from the top, sustained commitment throughout the organization, and an ability to translate challenge into opportunities. Firms must draw up clear plans of action and monitor progress closely.

Sustainability demands that we pay attention to the entire life cycles of our products and to the specific and changing needs of our customers.

Corporations that achieve ever more efficiency while preventing pollution through good housekeeping, materials substitution, cleaner technologies, and cleaner products and that strive for more efficient use and recovery of resources can be called "eco-efficient."

Long-term business-to-business partnerships and direct investment provide excellent opportunities to transfer the technology needed for sustainable development form those who have it to those who require it. This new concept of "technology cooperation" relies principally on private initiatives, but I can be greatly enhanced by support from governments and institutions engaged in overseas development work.

Farming and forestry, the businesses that sustain the livelihoods of almost half the world's population, are often influenced by market signals working against efficient resource use. Distortion farm subsidies should be removed to reflect the full costs of renewable resources. Farmers need access to clear property rights. Governments should improve the management of forests and water resources; this can often be achieved by providing the right market signals and regulations and by encouraging private ownership.

Many countries, both industrial and developing, could make much better use of the creative forces of local and international entrepreneurship by providing open and accessible markets, more streamlined regulatory systems with clear and equitably enforced rules, sound and transparent financial and legal systems, and efficient administration.

We cannot be absolutely sure of the extent of change needed in any area to meet the requirements of future generations. Human history is that of expanded supplies of renewable resources, substitution for limited ones, and ever greater efficiency in their use. We must move faster in these directions, assessing and adjusting as we learn more. This process will require substantial efforts in education and training, to increase awareness and encourage changes in lifestyles toward more sustainable forms of consumption.

A clear vision of a sustainable future mobilizes human energies to make the necessary changes, breaking out of familiar and established patterns. As leaders form all parts of society join forces in translating the vision into action, inertia is overcome and cooperation replaces confrontation.

We members of the BCSD commit ourselves to promote this new partnership in changing course toward our common future.

Policy Statement, International Federation of Consulting Engineers (FIDIC) – 1994[13]

There is a growing awareness that the earth cannot continue supporting increases in population and consumption. Mankind is threatening its own existence, in addition to that of many other forms of life, through global pollution and excessive consumption of limited resources.

Engineers have contributed to the quality of life through the provision of better water supplies and sanitation and by the development of natural resources, food, energy, and communication and transportation systems. These advancements have contributed to rapid population growth and environmental problems.

Consulting engineers accept the challenge of the endangered environment. Because of their professional training and background they have a particular role and obligation towards the protection of the environment. Engineers should provide leadership in achieving sustainable development, development that will meet the long term needs of future generations of all nations without causing major modification to the earth's ecosystems.

This role of the engineer should result in:

Careful evaluation of the environmental benefits and adverse impacts of proposed projects

Conservation of energy

Reduction in the use of non-renewable resources and increased re-use of materials

Reduced waste production through improved industrial processes, better transportation and distribution systems, and recycling of waste products

Sound agricultural and other land-management practices

Restoration or improvement of damaged land, polluted water supplies and disturbed ecosystems

Effective transfer of environmental knowledge and experience

---

[13] This policy statement was prepared by the International Federation of Consulting Engineers (FIDIC). More information is included in (FIDIC 1994), and is also available at: http://www.fidic.org. Reprinted with permission. Copyright © 1994 FIDIC. All rights reserved.

Ethics and Responsibilities

Worldwide steps are required to protect and improve our environment. The efforts must involve government, the public and the private sector.

Consulting engineers are trained and experienced in handling complex problems. They should combine their traditional skills with broader applications of physics, chemistry, biology and other disciplines to lead interdisciplinary teams directed at achieving acceptable environmental solutions.

Observing a code of conduct is a fundamental part of the profession of a consulting engineer. The goals of consulting engineers should include a commitment to achieve sustainable development. Consulting engineers should give highest priority to the short term and long-term welfare, health and safety of the community. They should consider regional, global and cumulative effects of projects in addition to local effects.

General Actions

FIDIC recommends that each consulting engineer should:

> Keep informed on global environmental trends and issues
>
> Discuss environmental problems with professionals from other disciplines
>
> Provide information to clients, the public and government about environmental problems and how adverse effects can be minimized
>
> Become involved in organizational activities, including assistance to governmental authorities, that promote the protection of the environment
>
> Encourage and promote appropriate environmental laws and regulations
>
> Actively support and participate in all forms of environmental education
>
> Promote research and development relevant to protecting and improving the environment

Project Actions

FIDIC recommends that consulting engineers should:

> Recommend that environmental studies be performed as part of all relevant projects. Such studies will normally require a multidisciplinary approach

Evaluate the positive and negative environmental impacts of each project. This evaluation might be based on a preliminary review of available information or on the engineer's experience. They should evaluate the basic functions and purposes behind a project. They should suggest alternatives to their clients if environmental risks emerge.

Develop improved approaches to environmental studies. Environmental effects should be considered early in the planning process. Studies should evaluate the long term consequences of environmental changes.

Makes clients aware that engineers can reduce but not always eliminate adverse environmental impacts. They legal and financial responsibilities of all parties should be clearly defined.

Urge clients to prevent or minimize the adverse environmental effects of projects in all phases; initial planning, design, construction, commissioning, operation, and decommissioning.

Finally, take appropriate action, or even decline to be associated with a project if the client is unwilling to support adequate efforts to evaluate the environmental issues or to mitigate environmental problems.

Policy of the Institution of Engineers, Australia (IEA) – 1997[14]

The Institution of Engineers, Australia (the Institution) requires that its members, in their practice of engineering, shall act in a manner that accelerates achievement of sustainability through:

Acknowledging that people are entitled to a healthy and productive life in harmony with nature;

Endeavoring to ensure that development today will not undermine the development and environment needs of present and future generations;

Recognizing that to achieve sustainability, environmental protection shall constitute an integral part of the development process, and that development cannot be considered in isolation from it;

Recognizing and taking into consideration the global environmental impacts of local actions and policies;

---

[14] This policy statement was prepared by the Institution of Engineers, Australia. More information is included in (IEA 1997), and is available at: http://www.ieaust.org.au/. Reprinted with permission of the publisher, Engineers Australia. Copyright © 1994 IEA. All rights reserved.

Using the precautionary approach to protect the environment and accepting that where there are threats of serious or irreversible damage, scientific uncertainty shall not be used to postpone cost-effective measure to prevent environmental degradation;

Reducing and eliminating unsustainable patterns of production and consumption;

Helping to ensure that environmental issues are handled with the participation of all concerned citizens;

Acknowledging that the community has a right to access, and an understand of, environmental information;

Promoting the internalization of environmental costs, taking into account he approach that the polluter should, in principle, bear the cost of pollution;

Supporting the transfer of knowledge and innovative technologies; and

Participating in a global partnership to conserve, protect and restore the health and integrity of the earth's ecosystem.

The Institution recognizes that there are important community principles, which will lead to sustainability. These global principles include that:

Eradication of poverty, the reduction of disparities in living standards, and the full participation of women, youth and indigenous people are essential to achieve sustainability;

People in developed countries bear a special responsibility to assist in the achievement of sustainability; and

Warfare is inherently destructive of sustainability and, in contrast, peace, development and environment protection are interdependent and indivisible.

In their practice of engineering, members of the Institution will be mindful of these principles and wherever possible use their engineering endeavors to progress sustainability.

Selected Principles of Sustainability

Many organizations have also developed statements on the broad principles, objectives and dimensions of sustainability. The following are provided as examples.

Guiding Principles of Sustainability. Institution of Engineers, Australia (IEA) – 1997[15]

Key guiding principles for managing a transition to sustainability are:

Precaution

The precautionary principles forms part of the Intergovernmental Agreement on the Environment (agreement among the States, Commonwealth and Local Government) and is enshrined in many international treaties affecting aspects of ecological sustainability. It states that where there are threats of serious or irreversible environmental damage, lack of full scientific certainty should not be used as a reason for postponing measures to prevent environmental degradation. Waiting until all the facts are known about a potential environmental threat may delay action until it is too late to reverse the conditions causing damage.

Equity for future generations

A fundamental tenet of sustainability is that future generations should be provided with at least the same economic, social and ecological opportunities as the generation making decisions today. This principle means that there is a need to preserve essential biological functions and processes. Ensuring equity between generations requires application of the precautionary principles because there is uncertainty about long term impacts of many activities.

Equity for all

This principle is taken to mean that all the people on earth at any one time have an equal right to a satisfactory quality of life. It recognizes that inequity cannot be the basis for long term success either economically or ecologically, inasmuch as those who have little stake in the system will not care for its resources. It also recognizes the inherent instability of a world in which there are gross disparities of opportunity among different groups of people.

The actual meaning of this principle in practice has not yet been very well defined. While governments in Australia can aim to ensure that their programs are equitable in providing access to disadvantaged groups, and can recognize and attempt to reduce Australian over consumption that perpetuates poverty in developing countries, it is not clear what will ensure equity for all people. Nevertheless, it is

---

[15] These principles were developed the Task Force for Sustainability of the Institution of Engineers, Australia. More information is available at: http://www.ieaust.org.au/.

important to take those actions within our power to reduce currently unacceptable levels of inequity.

The principle of environmental justice, which is an aspect of equity, is currently gaining increased attention. This principle is based on the need to rectify the widespread situation in which the major benefits of development flow to one group, while a different group experiences the most significant environmental damage. This applies to both global and local conditions. On a global scale, the greenhouse effect has resulted primarily from activities in industrialized countries, yet is predicted to create its most serious damage in poor island countries and the developing countries in Africa. At a local level, there is a need to change the practice of locating a large proportion of necessary but undesirable facilities, such as sewage treatment plants, landfills and hazardous industries, in disadvantaged areas.

Global responsibility

Local actions may produce effects well beyond their immediate area, even globally. Conditions such as the greenhouse effect and ozone depletion make clear that decisions taken locally cannot be made in isolation from their broader impacts. Governments in industrialized countries like Australia need to take account of the impact of past actions and continuing high levels of consumption on the potential fro people in less developed countries to achieve a decent quality of life.

Community involvement

People in all sectors of the community have a right to participate in and influence the economic, environmental, ecological and social decisions that affect them. The integration of economic, environmental and social issues requires decision-making to involve all sectors of the community.

Stewardship

Individuals, institution, and corporations need to take responsibility for the economic, social and environmental consequences of their actions.

Accountability

Governments are accountable to the communities they serve. This means that they have a responsibility to keep people informed. To communicate with communities, information needs to be provided in a form that people can understand, and opportunities have to be provided for communities to engage in two-way communication with Government.

Sustainability Dimensions and Objectives. The International Federation of Consulting Engineers (FIDIC) – 2000[16]

Environmental Dimension

Increase material efficiency by reducing the material demand of non-renewable goods.

Reduce the material intensity via substitution technologies

Enhance material recyclability

Reduce and control the use and dispersion of toxic materials

Reduce the energy required for transforming goods and supplying services

Support the instruments of international conventions and agreements

Maximize the sustainable use of biological and renewable resources

Consider the impact of planned projects on air, soil, water, flora, and fauna

Materials

Available resources do not allow the consumption of materials to continue to increase rapidly forever. The rate of increase must be reduced and eventually brought to the steady-state level. However, economic growth and production are expected to continue to increase sharply in the foreseeable future. In order to achieve stability, we must therefore reduce the net consumption of virgin materials per unit produced very significantly over the next few decades. This will demand many different measures, such as a large increase in recycling, techniques to reduce the materials' input for each unit of product of service generated, the development of new products and materials, and the development of ways to increase or optimize the durability of products.

Toxic Materials

This dispersion of toxic materials is generally believed to pose a major and increasing threat to nature and to human health. Focusing on the type and quantity of

---

[16] This sustainable development strategy was prepared by the International Federation of Consulting Engineers (FIDIC). More information is included in (FIDIC 1994), and is also available at: http://www.fidic.org. Reprinted with permission. Copyright © 1994 FIDIC. All rights reserved.

chemicals used in various materials is therefore very important for many types of projects. Mitigating measures include abandoning highly toxic chemicals by substituting less toxic substances, and the identification of alternative technological concepts that do not involve toxic chemicals.

Energy Requirements

Virtually all processes in today's industrialized society involve energy consumption. The global increase in energy consumption is one of the major challenges for sustainability. The most important energy consuming components of projects must be reviewed to see if there exists the potential to reduce energy requirements. There are two basic strategies for enhancing this potential: one involves increasing energy efficiency in every possible way; the other aims to introduce energy sources, which have reduced environmental impact.

Renewable Resources

It is generally believed that renewable resources, for example, biological resources, are more sustainable than non-renewable resources. However, this is only the case provided a resource is not over-exploited.

If over-exploitation takes place, the entire production system (ecosystem) may be destroyed – which may be much worse than the consequences of over-exploiting a non-renewable resource. So the careful management of all renewable resources is of the utmost importance.

Greenhouse Gases

The impact of human activity on the global climate is no longer questioned. Fossil fuel combustion resulting in carbon dioxide emissions is the main cause of the enhanced greenhouse effect leading to an overall increase in the global annual mean temperature. Contributing facts are agriculture and changes in land use, including deforestation, as well as certain industrial processes.

The Kyoto Protocol of 1997 established overall emission targets for the six principal greenhouse gases. So-called Kyoto mechanisms are being developed so that companies can trade the emission allowances, which are allocated by countries in accordance with the Protocol. Other Kyoto mechanisms are based on credits for earned emission abatement.

Recent assessments show that the secondary benefits of reducing greenhouse gas emissions (for instance, reduced air pollution) may also be extremely important. However, conventions on climate policy are coming into conflict with the increasingly deregulated power supply and gas industries.

Economic Dimension

Consider life-cycle costs

Internalize external costs

Consider alternative financing mechanisms

Develop appropriate economic instruments to promote sustainable consumption

Consider the economic impact on local structures

Life-cycle Costs

Costs as well as environmental issues must be considered from a life-cycle perspective. The reasons are clear-cut: a cost-saving measure in design and construction may increase significantly the cost of operation and maintenance, or reduce significantly the project lifetime. A realistic overall cost assessment for the client's benefit can only be made from a project life-cycle perspective. This approach is used infrequently today: it is also difficult to undertake in a precise and accurate way. However, there are no other alternatives for developing a sound basis for decisions.

Internalizing Costs

All major projects give rise to a large number and a broad variety of consequences for society, some of which will not affect the economic interest of owners or users while today's economic policies are in place. They may, however, affect the economic interest of other parties or stakeholders, as well as the environment. Such costs are termed external costs to the project. Generally speaking, they should be assessed since they can have major influence on the future of a project and on the return on investment. For example, they may increase stakeholder pressure for changing both the project design and local, regional or national policies, thus resulting in external costs being internalized in the project budget. Consequently, external costs, and the likely development of such costs, may have a major or even crucial influence on the long-term feasibility of a project.

Alternative Financing

Identifying alternative financing mechanisms is always an option for engineering projects. Life cycle costs assessments and the incorporation of external costs will make such considerations even more important in the future. This is because these life-cycle assessments, in addition to external costs, provide a broader picture of the overall project budget and its implications. Alternative financing models may then be viable, especially in a long-term perspective. For instance, assessments could result in a risk profile for a given project implying that public financing may be necessary because private investors could be reluctant to intervene.

Economic instruments

A variety of economic policy instruments may integrate a sustainability dimension into the economic decision making process. However at the present time they mostly refer to the environmental dimension:

> Environmental taxation (there is a trend towards comprehensive tax reform)
>
> Ecological tax reform (revenues from environmental taxes are being used to reduce taxes on labor)
>
> Sustainability asset management
>
> Subsidy reform (subsidies may have both damaging and beneficial impacts on sustainability)
>
> Extended cost-benefit analyses
>
> Tradeable permits/joint implementation
>
> "Green" procurement and "green" accounting
>
> Voluntary and negotiated agreements

In the long term, there may be a more radical shift away from taxing "goods" such as labor towards taxing "bads" such as environmental damage. The result of this shift to a combination of environmental taxes with a reduction in distortionary taxes may not only improve the environment but also yield positive economic benefits (the so-called "double dividend"). Although there is doubt that this dividend will materialize, it still makes sense to take advantage of the environmental and sustainability benefits of alternative tax structures, without counting on supposed economic benefits.

Social Dimension

> Enhance a participatory approach by involving stakeholders
>
> Promote public participation
>
> Promote the development of appropriate institutional frameworks

Consider the influence on the existing social framework

Assess the impact of health and the quality of life

Stakeholder Participation

A participatory approach is based on the premise that social issues and social acceptance cannot be analyzed and agreed upon in the traditional setting. Therefore, the only path to acceptance and consensus in controversial situations is the active involvement of the relevant stakeholders in the development, planning and implementation of a project. The involvement must be credible to the stakeholders so the process must be very proactive, transparent and fair. Such features will ensure that stakeholders do not feel they are being manipulated. The working procedures for participatory approach may be complicated, and demand a very professional methodology. However, the approach usually avoids conflicts and delays in the latter stages of a project. It also often leads to the introduction of new ideas for the design, development and implementation of a project.

Public Participation

In many cases, all the relevant key stakeholders cannot be identified at the start of a project. A selective participatory approach is inadequate, and the participation of a large audience is needed. For major, complex infrastructure projects, a dialogue with the general public is often necessary. This process involves general communication and information activities, contacts with the media, and the management of large public meetings. Once again, the dialogue must be proactive, transparent and fair, thus providing opportunities for genuinely interested parties to influence the outcome.

Institutional Framework

Addressing the long-term social changes in local or regional communities brought about by projects often calls for institutional development, either in the form of new institutions or via changes in existing institutions. For example, it is frequently necessary to train public administrators in the handling of participatory processes. In other cases, the creation of well-functioning non-governmental organizations may be a precondition for a "public opinion" to be articulated, and thus hopefully dealt with. Establishing institutions to support vulnerable groups of people may play an important role in preserving social stability.

Chapter IV
Bibliography

ABET (2003). *ABET Criteria for Accrediting Engineering Programs Effective for Evaluations During the 2003-2004 Accreditation Cycle*, available online at: http://www.abet.org/criteria.html

AIChE (2001). *Making EHS (Environmental, Health, and Safety) an Integral Part of Process Design*, American Institute of Chemical Engineers (AIChE)

ASCE (2001). *Policy on the Role of the Engineer in Sustainable Development*, American Society of Civil Engineers, Reston, VA, available online at: http://www.asce.org/pressroom/news/policy_details.cfm?hdlid=60

ASCE 2002. Code of Ethics http://www.asce.org/membership/codeofethics.cfm Accessed October 18, 2002.

ASEE (2002). *ASEE Statement on Sustainable Development Education*, American Society for Engineering Education (ASEE), available online at: http://www.asee.org/welcome/statements/sustain.cfm

ASTM (1994). *Standard Practice for Measuring Life Cycle Costs of Buildings and Building Systems*, ASTM International, E-917-94, West Conshohoken, PA.

ASTM (1995). *Standard Practice for Applying the Analytical Hierarchy Process to Multi Attribute Decision Analysis of Investments Related to Buildings and Building Systems*, ASTM International, E-1765-95, West Conshohoken, PA.

ATHENA (2004). ATHENA Sustainable Materials Institute Web Site: http://www.athenasmi.ca

BC Hydro (2004). BC Hydro: http://www.bchydro.com/wup/

Bordogna, J. (1998). *Tomorrow's Civil Systems Engineers – The Master Integrator*, Journal of Professional Issues in Engineering Education and Practice., ASCE, 124(2): 48-50.

CERF (1996). "Creating the 21st Century through Innovation; Engineering and Construction for Sustainable Development," Civil Engineering Research Foundation (CERF), CERF Report #96-5016E, Washington DC.

CNN (2004) Climate Neutral Network Web Site: http://www.climateneutral.com/

CWRT (2004) Center for Waste Reduction Technologies (CWRT) Web Page:
http://www.aiche.org/cwrt/index.htm

DESP (2004). District Energy St. Paul, Inc. Web Site: http://www.districtenergy.com/

Doyle, Randy, and Salmon, Jeff (2004). *Fort Hood's Buildings are Turning "Green,"* available online at:
http://www.cecer.army.mil/EARUpdate/NLFiles/2002/TurnGreen.cfm

DuPont (2004) Sustainable Growth 2003 Progress Report, DuPont, available online at:
http://www1.dupont.com/NASApp/dupontglobal/corp/index.jsp?page=/content/US/e n_US/social/SHE/usa/us1.html

EBN (2004). *Establishing Priorities with Green Buildings,* Environmental Building News Web Site: http://www.buildinggreen.com

EPA (2002). *Framework for Responsible Environmental Decision Making (FRED): Using Life Cycle Assessment to Evaluate Preferability of Products,* United States Environmental Protection Agency, EPA 660/R-00/095, by Science Applications International Corporation, Research Triangle Institute, and EcoSense, Inc.

Eureka (2004). Eureka Recycling Web Site: http://www.eurekarecycling.org/

FFC (2001). *Sustainable Federal Facilities: A Guide to Integrating Value Engineering, Life-Cycle Costing, and Sustainable Development,* Federal Facilities Council, Technical Report No. 142, National Academy Press, Washington, DC

Goodstein, E.S. (1995). *Economics and the Environment,* Prentice Hall, Englewood Cliffs, NJ, 575 pp.

Hatch, Henry J. (2002). *Sustainable Development,* excerpts from an address to The Presidents' Circle of the National Academies in November 14, 2002, Washington, DC

Hattis, David B., and Thomas E. Ware (1971). *The PBS Performance Specification For Office Buildings,* Report 10 527, The National Bureau of Standards, Washington CD. January 1971

Institution of Engineers, Australia – IEA (1997). *Towards Sustainable Engineering,* Engineers Australia – Task Force for Sustainability, Barton, Australia

Interface (2004). Interface Sustainability Web Site:
http://www.interfacesustainability.com/

International Federation of Consulting Engineers – FIDIC (1994). *Consulting Engineers and the Environment – FIDIC Guide for Actions*, International Federation of Consulting Engineers, Lausanne

International Federation of Consulting Engineers – FIDIC (1998). *Engineering Our Future – a FIDIC report*, International Federation of Consulting Engineers, Lausanne

Ft. Bragg (2004) *Sustainable Fort Bragg*, available online at: http://www.bragg.army.mil/sustainability/

ISO 14040 (1997). *Environmental Management – Life cycle Assessment – Principles and Framework*, International Standard 14040, International Standards Organization

ISO 14041 (1998). *Environmental Management – Life Cycle Assessment – Goal and Scope Definition and Inventory Analysis,* International Standard 14041, International Standards Organization

ISO 14042 (2000). *Environmental management – Life cycle Assessment – Life Cycle Impact Assessment,* International Standard 14042, International Standards Organization.

Kaplan, R.D. (2000). *The Coming Anarchy*, Vintage Books, New York, 198 pp.

Lippiatt, Barbara C. (2002). *BEES 3.0: Building for Environmental and Economic Sustainability; Technical Manual and User Guide,* NISTIR 6916, National Institute of Standards and Technology, available online at: http://www.bfrl.nist.gov/oae/software/bees.html

NAE (2002). *"Dialogue on the Engineers Role in Sustainable Development – Johannesburg and Beyond,"* held at the National Academy of Engineering in Washington, D.C. on June 24, 2002.

MDOC (2004). *Global Warming and Climate Change in Minnesota*, Minnesota Department of Commerce, Office of Environmental Assistance, and Pollution Control Agency, available online at: http://www.moea.state.mn.us/reduce/climatechange.cfm

NEC (2004). Saint Paul Neighborhood Energy Consortium (NEC) Web Site: http://www.spnec.org/

NRC (1999). *Our Common Journey: a Transition Toward Sustainability, Board on Sustainable Development*, National Research Council, National Academy Press.

Owens, J., et al., editors, (1997). *Life cycles Impact Assessment: the State-of-the-Art,* Society for Environmental Toxicology and Chemistry.

Roberts, D.V. (1994). "Sustainable Development – A Challenge for the Engineering Profession," in *The Role of Engineering in Sustainable Development*, American Association of Engineering, Societies, Washington, D.C., pp. 44-61.

Schmidheiny, S., Chase, R. and DeSimone, L. (1997). Signals of Change: Business Progress Towards Sustainable Development, World Business Council for Sustainable Development, Geneva

SPiRiT (2004). Sustainability Project Rating Tool (SpiRiT), U.S. Army Crops of Engineers, available online at: http://www.cccer.army.mil/SustDesign/SPiRit.cfm

U.S. Bureau of the Census (2002), *Statistical Abstract of the United States,* U.S. Bureau of the Census, Washington, DC, 2002

US Engineering Community (2002). "*A Declaration by the US Engineering Community to the World Summit on Sustainable Development,*" produced on June 24, 2002, at a meeting organized by the National Academy of Engineering, State Department, American Association of Engineering Societies, and American Institute of Chemical Engineers, in affiliation with the American Society of Civil Engineers, Engineers International Round Table, and World Federation of Engineering Organizations Committee on Technology.

USACE (2004) *Environmental Operating Principles*, U.S. Army Corps of Engineers, available online at: http://www.hq.usace.army.mil/cepa/envprinciples.htm

USGBC (2004). *Leadership in Energy and Environmental Design (LEED),* U.S. Green Building Council Web Site: http://www.usgbc.org

WBCSD (2001). *Mobility 2001, World Mobility at the End of the 21st Century and its Sustainability*, report prepared by the Massachusetts Institute of Technology and Charles River Associates for the Sustainability Working Group of the World Business Council for Sustainable Development, available online at: http://www.sustainablemobility.org/publications/mobility2001.asp

WBCSD (2004). Selected case studies on Innovation & Technology, Eco-Efficiency. Managing and Understanding Change, Dialogue and Partnership, Providing and Informing Customer Choice, Corporate Social Responsibility, and Creating Sustainable Livelihoods, available online at: http://www.wbcsd.org

WBDG (2004). *Whole Building Design Guide,* National Institute of Building Sciences, available online at: http://www.wbdg.org

WCED - World Commission on Environment and Development. (1987). *Our Common Future*. Oxford University Press, Washington, DC.

WFEO (2002). *Engineers and Sustainable Development*, Report prepared by the Committee on Technology (ComTech) of the World Federation of Engineering Organizations (WFEO), sponsored by the National Academy of Sciences and the National Science Foundation; CD developed by CH2M HILL.

Wilson, E. O. (2002). *The Future of Life*, Alfred A. Knopf, a division of Random House, Inc.,

WMO (2001). WMO Statement on the Status of the Global Climate in 2001, World Meteorological Organization, available online at: http://www.wmo.ch/web/Press/Press670.html

WSSD (2002) Official Web Site of the World Summit on Sustainable Development, held in Johannesburg, South Africa, in September, 2002: at: http://www.johannesburgsummit.org/

Xcel (2004). Xcel Energy Web Site: http://www.xcelenergy.com/XLWEB/CDA/

Appendix A
References, Resources, and Tools[17]

In recent years, numerous scholars from a wide range of disciplines have been contributing to a substantial and rapidly expanding academic body of knowledge on sustainable development. Consequently, it is not possible to list all relevant books and articles that have been published, or the many Internet sites where some of this material can be found. The intent of this appendix is to provide a sampling of references addressing the various dimensions of sustainability from multiple sources. They have been selected to provide the reader with an initial point of departure toward the understanding of the intellectual foundations of sustainability and sustainable development.

Selected References on the Various Dimensions of Sustainability

General References on Sustainable Development

Anderson, Ray C. (1998). *Mid-Course Correction, Toward a Sustainable Enterprise: The Interface Model.* Post Mills, VT: Chelsea Green

AtKisson, Alan (1999). *Believing Cassandra: An Optimist Looks at a Pessimist's World.* White River Junction, VT: Chelsea Green Publishing Company

Benyus, Janine M. (1997). *Biomimicry: Innovation Inspired by Nature.* New York: William and Morrow Company

Bartelmus, P. (1994). *Environment growth and development: The concepts and strategies of sustainability.* Routledge, New York, NY

Bennett, M. and James, P. (1998). *The Green Bottom Line: Environmental Accounting for Management.* Greenleaf Press, Sheffield, UK

---

[17] This appendix was prepared by Dr. Jorge Vanegas, Associate Professor of the School of Civil and Environmental Engineering at the Georgia Institute of Technology; with substantial contribution from Dr. Annie Pearce, Director of the Sustainable Facilities and Infrastructure Branch (SFI) of the Sustainable Facilities and Infrastructure (SFI) Branch of the Safety, Health, and Environmental Technology Division (SHETD), of the Electro-Optics, Environment and Materials Laboratory (EOEML) at the Georgia Tech Research Institute (GTRI).

Bennett, M. and James, P. (1999). *Sustainable Measures: Evaluation and Reporting of Environmental and Social Performance.* Greenleaf Press, Sheffield, UK

Clayton, A.M.H. and Radcliffe, N.J. (1996). *Sustainability: A Systems Approach.* Westview Press, Boulder, CO

Daly, Herman E. (1996). *Beyond Growth: The Economics of Sustainable Development.* Boston: Beacon Press

Daly, H.E. (1990). *"Towards Some Operational Principles of Sustainable Development,"* Ecological Economics, 2(1), 1-6

Daly, H.E., and Cobb, J.B., Jr. (1994). *For the Common Good,* 2nd ed. Beacon Press, Boston, MA

Doob, L.W. (1995). *Sustainers and Sustainability: Attitudes, Attributes, & Actions for Survival.* Praeger Publishing, Westport, CT

Frankel, Carl (1998). *In Earth's Company: Business, Environment and the Challenge of Sustainability.* Gabriola Island, British Columbia, Canada: New Society Publishers

Goodland, R., Daly, H.E., and El Serafy, S., eds. (1992). *Population, Technology, and Lifestyle: The Transition to Sustainability.* Island Press, Washington, DC

Hammond, Allen L. (1998). *Which World?: Scenarios for the 21st Century.* Washington, DC: World Resources Institute, Island Press

Harris, Jonathan (2000). *Rethinking Sustainability: Power, Knowledge, and Institutions.* Ann Arbor, Michigan: The University of Michigan Press

Hawken, Paul (1994). *The Ecology of Commerce: A Declaration of Sustainability.* New York: HarperCollins

Hawken, P., Lovins, A., and Lovins, L.H. (1999). *Natural Capitalism: Creating the Next Industrial Revolution.* Little, Brown, & Co., Boston, MA. Available free on the web at: http://www.naturalcapitalism.com

Howe, C.W. (1979). *Natural Resource Economics - Issues, Analysis and Policy.* John Wiley & Sons, New York, NY

Kinlaw, D. (1992). *Competitive and Green: Sustainable Performance in the Environmental Age.* Pfeiffer & Co., San Diego, CA

Lovins, Amory B.; Lovins, Hunter L.; von Weizsäcker, Ernst (1997). *Factor Four: Doubling Wealth, Halving Resource Use: The New Report to the Club of Rome.* London: Earthscan Publications

McDonough, W., and Braungart, M. (2002). *Cradle to Cradle – Remaking the Way We Make Things.* North Point Press, New York, NY

McKenzie-Mohr, D. and Smith, W. (1999). *Fostering Sustainable Behavior.* New Society Publishers, Gabriola Island, BC, Canada

Meadows, Donella; Meadows, Dennis L.; Randers, Jorgen (1992). *Beyond the Limits: Confronting Global Collapse/Envisioning a Sustainable Future: Executive Summary.* Post Mills, VT: Chelsea Green

Nattrass, B. and Altomare, M. (1999). *The Natural Step for Business: Wealth, Ecology, and the Evolutionary Corporation.* New Society Publishers, Gabriola Island, BC

Orr, David W. (1994). *Earth in Mind: On Education, Environment and the Human Prospect.* Washington, DC: Island Press

Pearce, D.W. and Warford, J.J. (1993). *World Without End.* Oxford University Press, Washington, DC

President's Council on Sustainable Development (1994). *Education for Sustainability: An Agenda for Action.* Washington: US Government Printing Office

Redefining Progress, Tyler Norris and Associates, and Sustainable Seattle (1997). *The Community Indicators Handbook: Measuring Progress Toward Healthy and Sustainable Communities.* San Francisco: Redefining Progress

Robèrt, K.H., Holmberg, J., and Eriksson, K.E. (1994). *"Socio-ecological Principles for a Sustainable Society - Scientific Background and Swedish Experience".* Ecological Economics

Robert, Karl-Henrik (1997). *The Natural Step to Sustainability.* Wingspread Journal, Spring 1997. Reprint

Roberts, D.V. (1994). *"Sustainable Development – A Challenge for the Engineering Profession,"* in Ellis, M.D., ed. The Role of Engineering in Sustainable Development. American Association of Engineering Societies, Washington, DC, 44-61

Rogers, P. (1991). "*The Economic Model*," in Chechile, R.A. and Carlisle, S., eds. Environmental Decision Making: A Multidisciplinary Perspective. Van Nostrand Reinhold, New York, NY

Wackernagel, Mathis; Rees, William (1996). *Our Ecological Footprint: Reducing Human Impact on the Earth*. The New Catalyst Bioregional Series. Philadelphia: New Society Publishers

Von Weizsacker, E., Lovins, A.B., and Lovins, L.H. (1998). *Factor Four: Doubling Wealth, Halving Resource Use*. EarthScan Publications, London, UK

General References on Sustainability of the Built Environment

Barnett, D.L. and Browning, W.D. (1995). *A Primer on Sustainable Building*. Rocky Mountain Institute, Snowmass, CO

Campbell, C. and Ogden, M. (1999). *Constructed Wetlands in the Sustainable Landscape*. John Wiley & Sons, Inc. New York, NY

Cole, Rick, Trish Kelly, and Judy Corbett, J. with Sharon Sprowls. (1998). *The Ahwahnee Principles for Smart Economic Development: An Implementation Guidebook*. Local Government Commission's Center for Livable Communities

Edwards, B., ed. (1998). *Green Buildings Pay*. E&FN Spon, London, UK

Ewing, R., DeAnna, M., Heflin, C. and Porter, D. (1996). *Best Development Practices: Doing the Right Thing and Making Money at the Same Time*. APA. Chicago, IL. Jain, R. K., Urban, L. V., Stacey, G. S., Balbach, H. E. (1993). *Environmental Assessment*. McGraw-Hill. New York, NY

Fiksel, J. (1996). *Design for Environment*. McGraw-Hill, New York, NY

Graedel, T.E. and Allenby, B.R. (1996). *Design for Environment*. Prentice Hall, Englewood Cliffs, NJ

Halliday, S.P. (1994). *Environmental Code of Practice for Buildings and their Services*. The Building Services Research and Information Association, Bracknell, Berkshire, UK

Krizek, Kevin and Joe Power. (1996). *A Planners Guide to Sustainable Development (Planners Advisory Service Report No.467)*. American Planning Association

Lewis, P. H., Jr. (1996). *Tomorrow By Design: A Regional Design Process for Sustainability.* John Wiley & Sons. New York, NY

Lyle, J. T. (1994). *Regenerative Design for Sustainable Development.* John Wiley & Sons. New York, NY

Mendler, S. and Odell, W. (2000). *The HOK Guide to Sustainable Design.* John Wiley & Sons, New York, NY

National Institute of Standards and Technology (2000). *Building for Environmental and Economic Sustainability (BEES) Version 2.0.* National Institute of Standards and Technology, Technology Administration, U.S. Department of Commerce, Gaithersburg, MD. Available for download at www.bfrl.nist.gov/oae/software/bees.html

PTI - Public Technology, Inc. (1996). *Sustainable Building Technical Manual: Green Building Design, Construction, and Operations.* Public Technology, Inc., Washington, DC. Available online at http://www.sustainable.doe.gov/freshstart/articles/ptipub.htm

Todd, N. J. and J. Todd. (1994). *From Eco-Cities to Living Machines: Principles of Ecological Design.* North Atlantic Books, Berkeley, CA

Van Der Ryn, S. and Calthorpe. (1986). *Sustainable Communities: A New Design Synthesis for Cities, Suburbs and Towns.* Sierra Club Books. San Francisco, CA

Wilson, A., Uncapher, J. L., McManigal, L., Lovins, L. H., Cureton, M., and Browning, W. D. (1998). *Green Development: Integrating Ecology and Real Estate.* Rocky Mountain Institute. John Wiley & Sons. New York, NY

Yeang, K.P. (1995). *Designing With Nature.* McGraw Hill, New York, NY

Sustainability Web Resources

Global Organizations to Promote Sustainability

Agenda 21 of the United Nations
http://www.un.org/esa/sustdev/agenda21.htm

Business Charter for Sustainable Development of the International Chamber of Commerce (ICC)
http://www.iccwbo.org/home/environment_and_energy/sdcharter/charter/about_charter/about_charter.asp

Coalition for Environmentally Responsible Economics (CERES)
http://www.ceres.org

Global Environment and Technology Foundation (GETF)
http://www.getf.org/homepage.cfm

International Institute for Sustainable Development (IISD)
http://www.iisd.org/

Investor Responsibility Research Center
http://www.irrc.org

ISO 14000 / ISO 14001 Environmental Management Guide
http://www.iso14000-iso14001-environmental-management.com/

Leadership for Environment and Development (LEAD) International
http://www.lead.org

My Earth
http://www.myearth.org/

SD Gateway
http://www.sdgateway.net/default.htm

Sustainable Construction Program of the International Council for Research and Innovation in Building and Construction (CIB)
http://www.cibworld.nl/4DCGI!index.shtml?RSES=2004921171465681

World Business Council for Sustainable Development (WBCSD)
http://www.wbcsd.org

Worldwatch Institute
http://www.worldwatch.org/

Worldwatch Resources Institute
http://www.wri.org/

Comprehensive Sustainability Sites

A Guide to Internet Resources in Sustainable Development
http://www.caf.wvu.edu/gdsouzawww/guide.html
Comprehensive listing of links to sustainability resources compiled by
Gerard E. D'Souza, West Virginia University.

Argus Clearinghouse: Sustainable Development
http://www.clearinghouse.net/cgi-
bin/chadmin/viewcat/Environment/sustainable_development?kywd
An internet research library with links to sustainability resources.

Global Environmental Options - GEO Link Library
http://www.geonetwork.org/links/index.html
Links to over 500 informative sites.

Solstice/Center for Renewable Energy and Sustainable Technology
http://www.solstice.crest.org/sustainable/index.shtml
On-line resource for sustainable energy.

Sustainability Web Ring
http://www.sdgateway.net/webring/default.htm
An Internet tool that allows users to navigate easily between Web
sites that address with the principles, policies, and best practices for
sustainable development.

The World Wide Web Virtual Library: Sustainable Development
http://www.ulb.ac.be/ceese/meta/sustvl.html
A comprehensive list of internet sites dealing with sustainable
development.

Selected Private Sector and Nonprofit Sustainability Web Resources

Advanced Building Technologies
http://www.advancedbuildings.org/
A British resource describing over 90 best practices for commercial green
buildings.

BuildingGreen.com
http://www.buildinggreen.com/
Home of the resources associated with Environmental Building News.
Excellent subscription-based resources, as well as free archives of some of
the best articles from EBN on green building.

Envirolink
http://www.envirolink.org/
An extensive database of sustainability-related resources organized by topic, actions you can take, resources, publications, jobs, and other useful indexing mechanisms.

GreenBiz.com
http://www.greenbiz.com/
Home of GreenBiz.com, one of the best resources for businesses going green. Includes a green job clearinghouse, online reference desk, a weekly newsletter that's concise and informative, and many other resources.

Guide to Resource Efficient Building Elements
http://www.crbt.org/index.html
Link to the Center for Resourceful Building Technology's online version of this guide for residential sustainable technologies.

Minnesota Sustainable Design Guide
http://www.sustainabledesignguide.umn.edu/
An excellent design guide organized by topic and life cycle phase to provide easy navigation of appropriate sustainable building practices.

Sustainable Behavior Web Site
http://www.cbsm.com/
Link to resources on community-based social marketing, a technique that can be used to design sustainability programs for success.

Sustainable Sources
http://www.greenbuilder.com/general/BuildingSources.html
Includes online directory of green building professionals, along with a frequently updated calendar where the best events are posted.

Selected Federal Sustainability Web Resources

DOE Funding Page
http://www.eere.energy.gov/buildings/solicitations.cfm
This page describes opportunities for partnering with the Department of Energy for funding green building projects. It also links to a list of funding opportunities updated monthly.

DOE High Performance Buildings Page
http://www.eere.energy.gov/buildings/highperformance/
Links to DOE's high performance building resources, including one of the most detailed sets of case studies available, an online toolbox, and performance metrics for measuring building performance.

DOE Smart Communities Network
http://www.sustainable.doe.gov/
DOE's site for topics relating to sustainable planning and communities. Includes an online toolkit of sustainable development resources.

Energy Star Home Page
http://www.energystar.gov/
Link to energy star product listings, rating criteria, and other useful information.

FEMP Software Tools
http://www.eere.energy.gov/femp/techassist/softwaretools/softwaretools.html
Link to free, downloadable software for energy and water analysis of buildings. Also includes software for life cycle cost analysis.

Greening Federal Facilities Guide
http://www.eere.energy.gov/femp/techassist/green_fed_facilities.html
Link to the second edition of Greening Federal Facilities, a downloadable guide highlighting practical and pilot tested sustainability best practices for federal buildings.

GSA's Real Property Sustainable Development Guide
http://policyworks.gov/org/main/mp/gsa/home.html
Good introductory document to explain the principles of sustainable design for real property.

National Park Service Guiding Principles of Sustainable Design
http://www.nps.gov/dsc/dsgncnstr/gpsd/
Link to NPS's excellent guide on sustainable design principles that covers the whole life cycle of sustainable facilities.

Pacific Northwest Laboratory (PNL) Sustainable Design Page
http://www.pnl.gov/docsustainabledesign/index.html
Variety of resources to support sustainable design, including the EDGE software for sustainable design.

USDA Forest Service Built Environment Image Guide
http://www.fs.fed.us/recreation/programs/beig/
Link to the Forest Service's guide for regionally appropriate, sustainable buildings. Excellent introduction to regional vernacular architecture for the U.S.

Whole Building Design Guide
http://www.wbdg.org/

One stop shop for the building professional organized by building type, design objectives, and product type. This guide links to other key sites on the web and contains a current set of links to all official agency policies on sustainable design.

Selected Military Sustainability Web Resources

AFCEE's Sustainable Design Web Site
http://www.afcce.brooks.af.mil/eq/programs/progpage.asp?PID=27
AFCEE's online resources for sustainability. Includes USAF case studies and real SOW's, link to the Environmentally Responsible Facilities Guide, and other important sustainability resources.

Air Force Center for Environmental Excellence (AFCEE) Sustainable Design Toolbox
http://www.afcee.brooks.af.mil/green/resources/toolbox/TOOLBOX.asp
Link to resources useful at an installation level for meeting sustainability goals.

Air Force Sustainable Facilities Guide
http://www.afcee.brooks.af.mil/dc/dcd/arch/rfg/index.html
Direct link to AFCEE's new tool for sustainable facilities, based on the LEED Green Building Rating System.

Resources on Green Guide Specs
http://www.afcee.brooks.af.mil/green/resources/ufgs.asp
Central location to keep up with the greening of USACE and NAVFAC guide specs

Sustainable Design and Development Resource of the Construction Engineering Research Laboratory (CERL) of the Engineer Research and development Center (ERDC) of the U.S. Army Corps of Engineers (USACE)
http://www.cecer.army.mil/sustdesign/
The absolute best site to keep track of what is going on in the federal government. Regularly updated, this site covers not only the best military but also the best federal resources.

USACE Ideas for Sustainable Buildings
http://www.cecer.army.mil/sustdesign/Discussion.cfm
List of ideas for creating more sustainable facilities developed by the Corps of Engineers (good ideas across a variety of spectra).

USACE Sustainable Design Web Site
http://www.hq.usace.army.mil/isd/librarie/RP/Sustainability/Sustainable%

20Design%20&%20Development.htm
Link to the Corps of Engineers' list of sustainable design resources.

Sustainable Building Web Sites

Bion Howard's Green Building Primer
http://www.nrg-builder.com/greenbld.htm
This primer on green building is an excellent starting point for learning
industry-determined requirements and characteristics of green buildings.

Building Energy Software Tools
http://www.eren.doe.gov/buildings/tools_directory/
Described here are 211 energy-related software tools for buildings, with
an emphasis on using renewable energy and achieving energy efficiency
and sustainability in buildings.

Building for Environmental and Economic Sustainability (BEES) System
http://www.bfrl.nist.gov/oae/software/bees.html
Evaluates the relative environmental and economic impacts of building
materials.

Building Technology Program of the Office of Energy Efficient and Renewable Energy of the U.S. Department of Energy
http://www.eere.energy.gov/buildings/index.cfm?flash=yes
Building information for homeowners; commercial owners; builders and
designers; researchers, and public officials.

City of Austin Green Building Site
http://www.greenbuilder.com
Links to most of the happenings in the green building world. The main
feature is the City of Austin's Green Building Guide, featuring extensive
knowledge and design advice in the primary areas of water, energy,
materials, and waste.

Environmental Building News Recommended Readings
http://www.buildinggreen.com/index.cfm
The leading newsletter on environmentally responsible design and
construction. Recommended readings on green and sustainable materials.

Georgia Tech's Sustainable Facilities & Infrastructure Program
http://maven.gtri.gatech.edu/sfi
Contains selected publications, information about courses and training
opportunities, databases, and links to the best green building sites.

Green Building Advisor
http://solstice.crest.org/software-central/gba/
A software program directed at developing healthy and producing indoor spaces, while reducing environmental impacts of building projects.

Green Buildings Center of Excellence for Sustainable Development of the U.S. Department of Energy
http://www.sustainable.doe.gov/buildings/gbintro.html
Resources include explanations of building principles; building programs; rating systems; codes and ordinances; green building efforts and successes in the U.S.; and publications and educational materials.

Leadership in Energy and Environmental Design (LEED) System of the U.S. Green Building Council
http://www.usgbc.org/programs/index.htm
Evaluates environmental performance from a "whole building" perspective over a building's life cycle, and is intended to be a definitive standard for what constitutes a "green building".

Oikos® Green Building Source
http://oikos.com/
A variety of links and resources focusing on applied aspects of green building, including a link to an electronic bookstore which features some of the best titles on sustainable construction.

Sustainable Buildings Industry Council (SBIC)
http://www.sbicouncil.org/home/index.html
Nonprofit organization to advance the design, affordability, energy performance, and environmental soundness of residential, institutional, and commercial buildings nationwide

Sustainable Information Directory of the U.S. National Park Service
http://www.nps.gov/sustain/
Searchable site with information on sustainable communities, facilities, and products.

U.S. Green Building Council
http://www.usgbc.org
A nonprofit, consensus coalition to promote green building.

Appendix B
ASCE Committee on Sustainability

Albert A. Grant, Chair

Jorge A. Vanegas, Editor

Henry J. Hatch

Miriam Heller

William E. Kelly

Michael R. Sanio

Howard A. Schirmer, Jr.

Richard N. Wright

Index